STERLING
Education

Essential Chemistry

Chemical Kinetics
&
Equilibrium

4th edition

4   3   2   1

ISBN-13: 979-8-8855706-3-3

Sterling Education materials are available at quantity discounts.

Contact info@sterling–prep.com

Sterling Education
6 Liberty Square #11
Boston, MA 02109

Published by Sterling Education

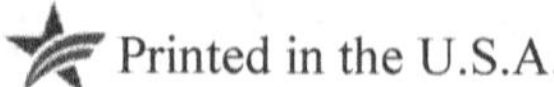 Printed in the U.S.A.

STERLING
Education

From the foundations of chemical reactions to the complex mechanisms of atomic particles, *Essential Chemistry Self-Teaching Guides* are a comprehensive compendium of clearly explained texts to learn and master these multifaceted chemistry topics.

These guides provide a detailed review of the fundamental mechanisms of chemical and physical processes at the atomic level. Develop a better understanding of the electronic structure of elements, principles of chemical bonding, phases of matter, types and mechanisms of chemical reactions, and principles of solutions and acid-base equilibria. Learn about rate processes in chemical reactions, empirical and molecular formulas, enthalpy, entropy, oxidation number, the laws of thermodynamics, and electrochemistry. Reinforce your learning by working through the practice questions and step-by-step solutions.

Created by highly qualified chemistry instructors, researchers, and education specialists, these books empower readers by helping them increase their understanding of general chemistry.

We sincerely hope that these guides are valuable for your learning.

250909akp

Featured on

**Essential Chemistry Self-Teaching Guides**

Electronic Structure & Periodic Table

Chemical Bonding

States of Matter & Phase Equilibria

Stoichiometry

Solution Chemistry

Chemical Kinetics & Equilibrium

Acids & Bases

Chemical Thermodynamics

Electrochemistry

Visit our Amazon store

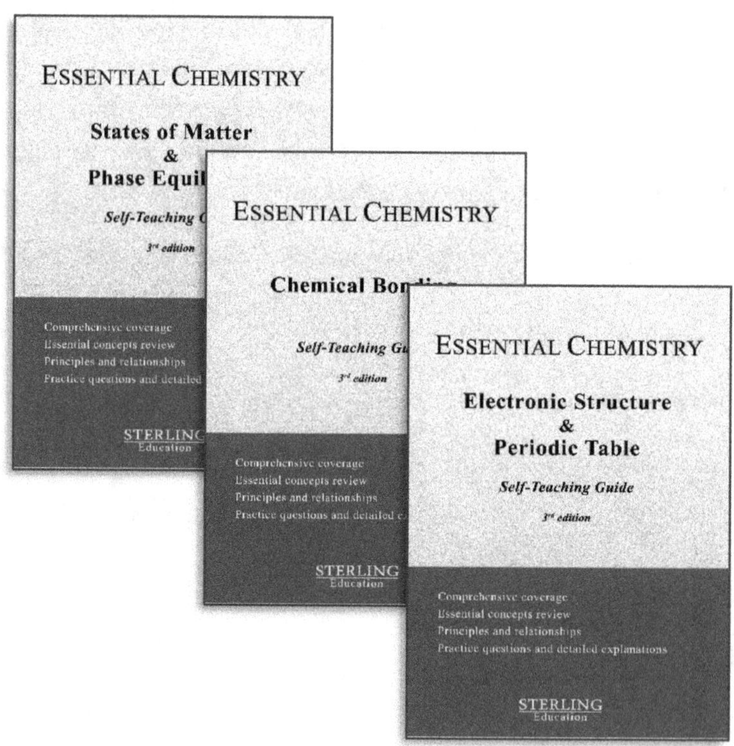

Essential Physics Self-Teaching Guides

Kinematics and Dynamics

Equilibrium and Momentum

Force, Motion and Gravitation

Work and Energy

Fluids and Solids

Waves and Periodic Motion

Light and Optics

Sound

Electrostatics and Electromagnetism

Electric Circuits

Thermal Physics

Atomic and Nuclear Physics

Visit our Amazon store

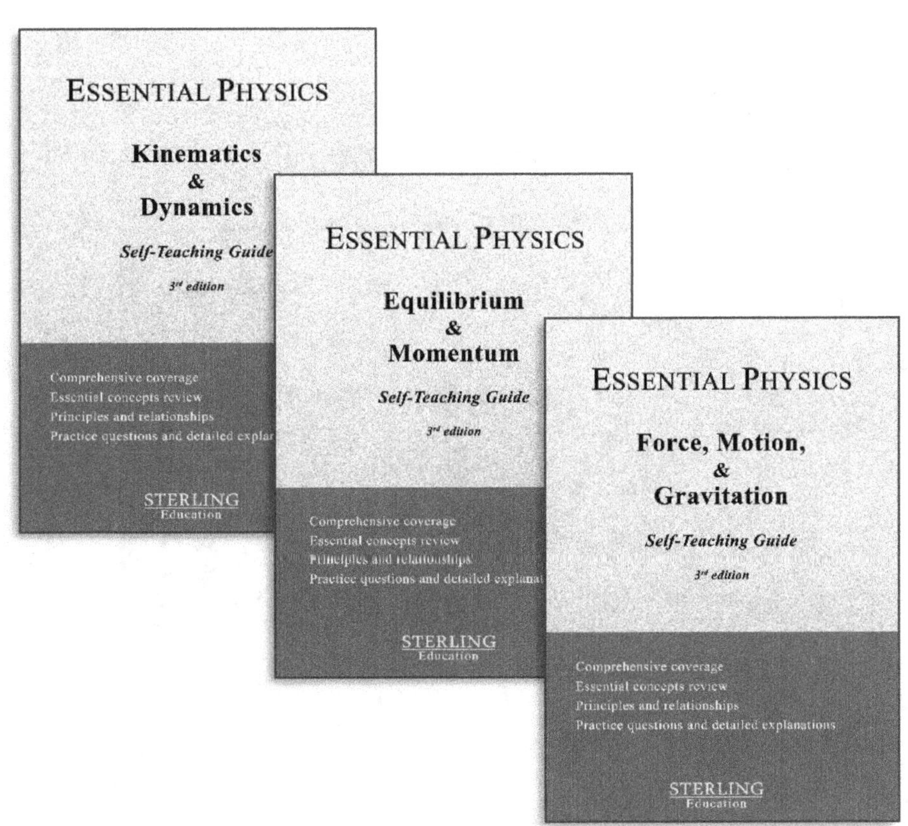

Eukaryotic Cell & Cellular Metabolism

Molecular Biology & Genetics

Nervous & Endocrine Systems

Circulatory, Respiratory & Immune Systems

Digestive & Excretory Systems

Muscle, Skeletal & Integumentary Systems

Reproduction & Development

Microbiology

Plants & Photosynthesis

Evolution, Classification & Diversity

Ecology & Population Biology

Visit our Amazon store

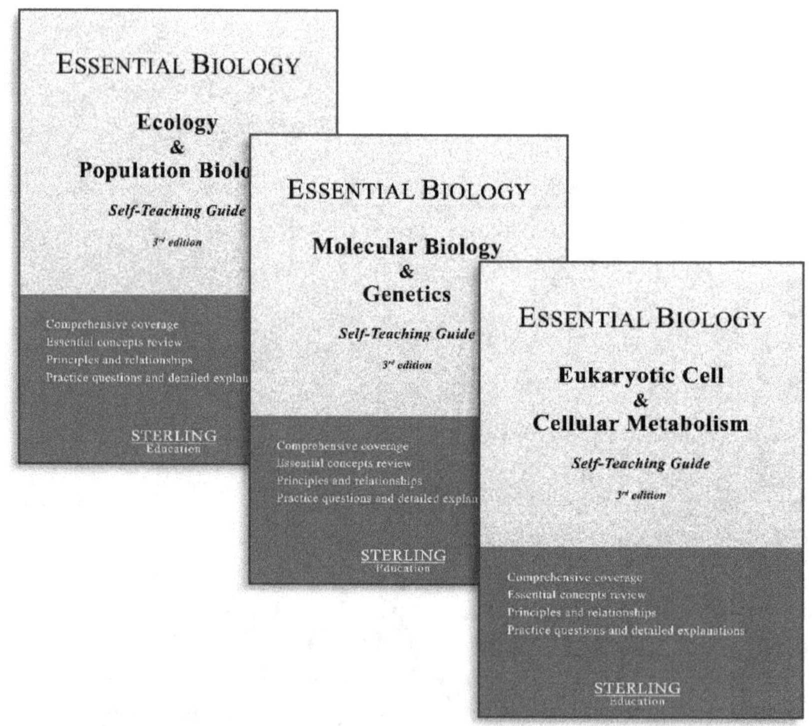

Everything You Always Wanted to Know About…

Chemistry

Physics

Cell & Molecular Biology

Organismal Biology

Human Anatomy & Physiology

Environmental Science

American History

American Law

American Government & Politics

Comparative Government & Politics

World History

European History

Psychology

Sociology

Human Geography

Visit our Amazon store

For online practice resources visit

https://www.sterling-prep.com

Table of Contents

Table of Contents (*continued*)

Table of Contents (*continued*)

Table of Contents (*continued*)

REVIEW

Chemical Kinetics & Equilibrium

Page intentionally left blank

The Nature of Science

Scientific discovery

Observation is the first step toward scientific discovery. Observations recognize patterns and anomalies in unexplained phenomena.

Hypothesis is advanced as a proposed explanation for the observation.

Evaluating a hypothesis involves experiments and additional observations (i.e., recording and collecting data).

Experimental observations are made while manipulating *independent variables* (e.g., how long the plant is exposed to sunlight per twenty-four hours) with *dependent variables* (data collected regarding height measurements).

Observations generate *data* that:

    1) *support the hypothesis* (i.e., different from proves) or

    2) *refute the hypothesis*.

Hypothesis are modified by narrowing or expanding their application to explain additional data. More experiments are performed, and the results either *support* (not prove) or *refute* the hypothesis during this iterative process.

Principles are the initially proposed explanations that are specific and apply to a narrow range of phenomena.

Theories build upon valid hypotheses

Theories are proposed to account for the observations when a hypothesis is valid over a range (e.g., sunlight, temperature, or humidity).

Theories are advanced to explain the phenomena and to predict what will happen under other conditions with different variables.

Further observations determine whether the predictions were accurate.

Research (re-search or *again search*) continues as these many observations support hypotheses and theories.

When theories are accurate over various scenarios and experiments, *Laws* are developed based on these repeatedly replicated results.

Laws as established theories

Laws are brief descriptions of how nature behaves in a broad set of circumstances. They are the consolidation of several theories. Laws are the products of multiple theories tested and proven, culminating in a broad claim regarding those predictions.

Laws are widely applicable explanations. Laws with broad and straightforward claims are more robust than those that make claims on a narrower, more specific set of variables and parameters.

The number of Laws describing physical phenomena is minuscule compared to the number of theories and magnitudes, which are less than the number of supported hypotheses.

The number of refuted (unsupported) hypotheses may be several magnitudes larger.

Failed experiments do not typically refer to the physical process where the researcher made a human error during the experiment. A failed experiment means that the *data does not support the hypothesis.* Therefore, after additional trials to validate the "failed" results, the hypothesis must be modified to predict the results of either future experiments or objective observations in the physical world.

Theories, models and laws

Theories emerge from a hypothesis that experimental data has repeatedly supported. Theories are detailed statements that provide testable predictions of the behavior of natural phenomena.

Theories are established to explain observations and then tested (supported or refuted) based on their predictions. Tested theories can articulate the boundaries of the tested hypothesis.

Theories are more specific than *hypotheses* but vaguer and more encompassing than *models*.

Models are constructed to explain the observed behavior if a hypothesis accurately predicts a phenomenon. They are more specific in their application than theories.

Data collected to develop the theories provide the conceptual foundation for constructing models. The model creates mental pictures (e.g., complicated weather patterns displayed as visual pictures for the viewer) of the physical phenomena.

Models should be consistent with scientific theories' predictions and support a comprehensive understanding of a phenomenon. Models allow for predictive outcomes under specific theories (e.g., when the wind speed reaches x, then y will occur).

Multiple models are integrated to explain a portion, or the entire scope, of a *theory*.

Understand the limitations of a model and do not apply it dogmatically unless its applicability is evaluated. A model is an underlying representation of a part of a theory; it is not intended to provide a complete picture of every phenomenon under that theory.

Instead, models provide the fundamental aspects of the theory.

Reaction Rates

Measuring reaction rate

Reaction rate is the rate of change in the concentration of reactants or products (i.e., how fast a reactant is consumed and how fast a product is formed).

Reaction rates:

$$\text{rate of reaction} = \frac{\text{change in concentration of reactant or product}}{\text{change in time}}$$

or

$$\text{rate of reaction} = \frac{\Delta\ \text{concentration}}{\Delta\ \text{time}}$$

Rate is concentration divided by time; the unit is molarity per second (M/s).

Expressing reaction rates

Rate of reaction is written for both the forward and reverse directions of chemical reactions.

Subscript "fwd" or "rev" specifies direction of the reaction rate.

Forward rate of reaction:

$$\text{rate}_{\text{fwd}} = \frac{-\Delta\text{reactant}}{\Delta\text{time}} \qquad \textit{reactant disappears}$$

$$\text{rate}_{\text{fwd}} = \frac{\Delta\text{product}}{\Delta\text{time}} \qquad \textit{product forms}$$

Reverse rate of reaction:

$$\text{rate}_{\text{rev}} = \frac{-\Delta\text{product}}{\Delta\text{time}} \qquad \textit{product disappears}$$

$$\text{rate}_{\text{rev}} = \frac{\Delta\text{reactant}}{\Delta\text{time}} \qquad \textit{reactant reforms}$$

Collision theory

A reaction between two molecules occurs *if* the molecules:

1) have sufficient *kinetic energy,* and

2) molecules are *appropriately oriented* to start a reaction in the image below.

Collision theory focuses on gas-phase chemical reactions.

Three *collision-related* factors affect the rate of a chemical reaction:

1. *Collision **frequency**:* Increasing the frequency of reactant collisions increases the reaction rate. More collisions increase the probability that a collision produces the product.

2. *Collision **energy**:* Reactants must collide with enough energy to *form new bonds* for a reaction to occur.

3. *Collision **orientation**:* Reactants must have the correct orientation (i.e., be properly aligned) for products to form.

Reaction forms a product from two molecules if the molecules have:

1) sufficient kinetic energy

2) appropriately oriented molecules

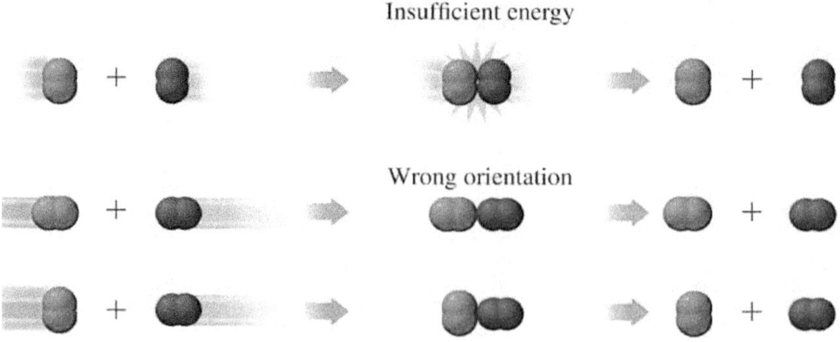

The two requirements of sufficient kinetic energy and appropriately oriented molecules are satisfied when $N_2 + O_2$ transforms into NO

Collisions that do not form products

Rate Dependence on Concentration of Reactants

Rate law

Rate law expression describes how the reaction rate changes with concentrations of reactants and products.

For example, consider the general expression for a homogeneous reaction:

$$a\,A + b\,B \longrightarrow c\,C + d\,D$$

Rate law expression is:

$$rate = -\Delta[A] \,/\, \Delta t$$

$$-\Delta[A] \,/\, \Delta t = k[A]^{x}{\cdot}[B]^{y}$$

where $[A]$ and $[B]$ are initial concentrations of the reactants, x and y are partial rate orders of the reaction determined experimentally, and k is the rate constant.

Rate constant k

Rate constant (k) is *empirically* derived and quantifies a chemical reaction. It accounts for factors influencing reaction rate, including *temperature* and *solvent*.

For example, consider the general expression for a homogeneous reaction:

$$a\,A + b\,B \longrightarrow c\,C + d\,D$$

Rate constant (k) expression is:

$$rate = k(T)[A]^{x}{\cdot}[B]^{y}$$

$$k(T) = [A]^{x}{\cdot}[B]^{y} \,/\, rate$$

where $[A]$ and $[B]$ are initial concentrations of the reactants, x and y are partial rate orders of the reaction determined experimentally, and $k(T)$ is the rate constant dependent on the reaction mechanism.

Therefore, the constant is specific to the *experimental conditions* (i.e., *empirically* derived).

There is *no correlation* between the stoichiometric coefficients (a, b) and the rate exponents (x, y).

Empirically derived exponents are often small integers, fractions, or zero.

Relative reactant and product concentrations

If the reactants' initial concentration increases, there are more reactant molecules; thus, molecules are closer and collide more frequently.

Higher collision frequency results in a *higher rate* of reaction.

During reactions, concentration of *reactant* molecules decreases, and *forward rate decreases*.

Thus, the concentration of *product* molecules increases, and the *reverse reaction rate increases*.

Partial rate order

Partial rate order is the exponent for each concentration term.

- For reactions ***zero-order*** for a reactant, the rate is constant and does *not* depend on the reactant's concentration.

  Reactant's exponent is zero.

  Thus, the term equals one (since anything to the zeroth power is one).

  Since it equals one, it is excluded from the rate law expression, as the reaction rate does not depend on a zero-order reactant.

- For reactions ***first-order*** for a reactant, the reaction rate is directly proportional to the concentration of that reactant.

  The exponent is one.

- For reactions ***second-order*** for a reactant, the reaction rate is directly proportional to the square of the concentration of that reactant.

  The exponent is two.

Differential rate laws

Overall order of the reaction is the sum of the partial orders.

Most differential rate laws are zero-, first-, or second-order overall:

- *Zero-order overall*:

  The differential rate law is generally $r = k$

  The units of the rate constant k are mole $\times$ L^{-1} $\times$ sec^{-1} (or M/s)

- *First-order overall*:

  The differential rate law is generally $r = k[A]$

  The units of the rate constant k are sec^{-1}

- *Second-order overall*:

  The differential rate law is generally $r = k[A]^2$

  The units of the rate constant k are L $\times$ mole^{-1} $\times$ sec^{-1} (or M$^{-1} \cdot$ s^{-1})

Notes for active learning

Reaction Order

Overall order of reaction

Example below illustrates how to sum partial orders to obtain an equation's overall order.

For example, consider the reaction:

$$2\,NO\,(g) + O_2\,(g) \leftrightarrow 2\,NO_2\,(g)$$

Rate law was experimentally determined to be:

$$rate\ law = k[NO]^2 \cdot [O_2]$$

For rate law, the partial order for NO is 2, and the partial order for O_2 is 1.

The overall order is three; this reaction is *third order overall*.

Note that the orders of the reactions match the reaction coefficients, but rate orders and coefficients are *not necessarily* correlated. Coefficients only equal rate orders for elementary-step reactions. It is often unknown whether the reaction is an elementary or multi-step reaction, unless it is explicitly stated.

Molecularity

Molecularity is important in defining the overall order of a reaction.

Reaction mechanism's molecularity is determined by the number of molecules (typically the coefficients) or ions participating in the rate-determining step.

Unimolecular reactions have a single reactant in the transition state.

Bimolecular reactions have a mechanism involving two reacting species.

Termolecular reactions have a mechanism involving three reacting species; they are rare because the necessary factors are so specific that they are unlikely to occur.

Reaction Type	Overall Reaction Order	Rate Law(s)
Unimolecular	1	$r = k[A]$
Bimolecular	2	$r = k[A]^2, r = k[A]\cdot[B]$
Termolecular	3	$r = k[A]^3, r = k[A]^2\cdot[B], r = k[A]\cdot[B]\cdot[C]$
Zero order	0	$r = k$

Differential *vs.* integrated rate law

Differential rate law and *integrated rate law* express specific rate laws.

Differential rate laws (above) express the rate of reaction *vs.* concentration.

Integrated rate laws express the change in concentration of components *vs.* time.

Integrated rate laws are:

Integrated rate law for 0th order

$$[A]_t = -kt + [A]_0$$

Integrated rate law for 1st order

$$\ln[A]_t = -kt + \ln[A]_0$$

Integrated rate law for 2nd order

$$1 / [A] = 1 / [A]_0 + kt$$

Initial experimental rate method

Method of initial rates determines the rate law by using experimental data to determine the rate order of reactants.

For example, experiment results show reaction rates relative to initial concentrations of reactants:

Concentrations (M)

$[O_2]$		Rate (M/s)	initial $[NO]$
0.10	Trial 1	1.2×10^{-8}	0.10
0.20	Trial 2	2.4×10^{-8}	0.10
0.10	Trial 3	1.08×10^{-7}	0.30

Based on the data above, determine:

a) reaction orders of the reactants

b) value and units of k

Rate law by concentrations

To determine the rate law, compare two trials in which the concentration of one reactant changes while the other reactant's concentration is held constant.

It is necessary to determine the reactant's partial order for changed concentrations.

Keep the other reactant concentration constant, so the change in rate is due to one reactant's concentration change.

Consider the results for Trial 1 and Trial 2.

Comparing these two trials reveals that the NO concentration remains constant, while the O_2 concentration is doubled.

To determine the partial order of O_2, note that doubling the concentration of O_2 doubles the reaction rate.

Expressed as:

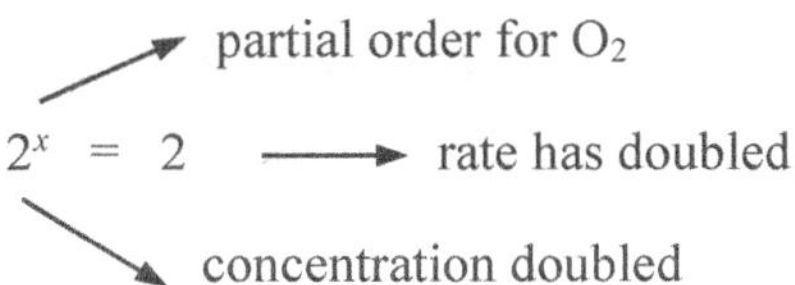

For equality, x equals one.

Therefore, the partial order for O_2 is first-order.

To determine the partial order of NO, select two trials with the concentration of O_2 constant while [NO] varies.

These criteria are satisfied with Trial 1 and Trial 3.

Concentration of O_2 is held constant, while the concentration of NO has tripled.

Under these conditions, the reaction rate increased ninefold.

Expressed as:

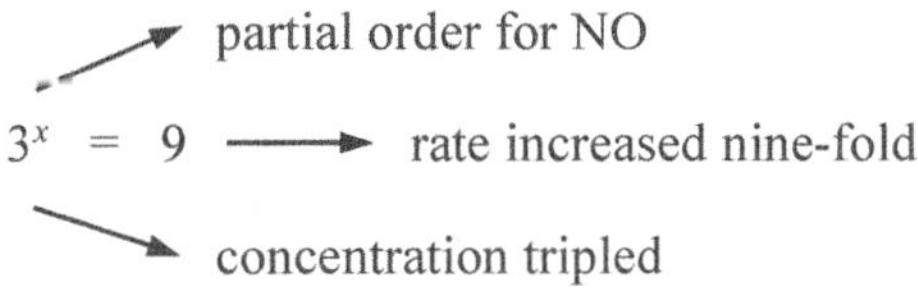

For equality, x equals two.

Therefore, the partial order for NO is second-order.

Calculating rate order experimentally

The calculated rate orders are substituted into the rate law:

$$rate\ law = k[A]^x \cdot [B]^y$$

$$rate\ law = k[NO]^2 \cdot [O_2]$$

Partial first orders are typically omitted, so $[O_2]$ does not have an exponent; exponent is assumed as one.

With partial rate orders, it is possible to solve the rate constant, k.

For example, choose an experimental trial, and substitute the values of *rate* and *concentrations* of NO and O_2.

Using data from Trial 1:

$$rate = k[NO]^2 \cdot [O_2]$$

$$k = rate\ /\ [NO]^2 \cdot [O_2]$$

$$k = 1.2 \times 10^{-8}\ M^{-2} \cdot s^{-1}\ /\ [(0.10\ M)^2 \cdot (0.10\ M)]$$

$$k = 1.2 \times 10^{-5}\ M^{-2} \cdot s^{-1}$$

Thus, the rate law becomes:

$$rate = 1.2 \times 10^{-5}\ M^{-2} \cdot s^{-1}[NO]^2 \cdot [O_2]$$

The reaction is third order overall, and the units are $M^{-2} \times s^{-1}$.

Notice that the units of k cancel to yield M/s for the reaction rate.

Activation energy

The reactants must have sufficient energy to *break existing bonds* and *form new ones* for a reaction to proceed.

This minimum energy threshold is the *activation energy* (E_a).

Activation energy barrier (E_a, represented as the hill on the reaction profile energy vs. reaction progress graph) is the difference between the energy of the reactants and the highest point on the reaction profile toward product formation.

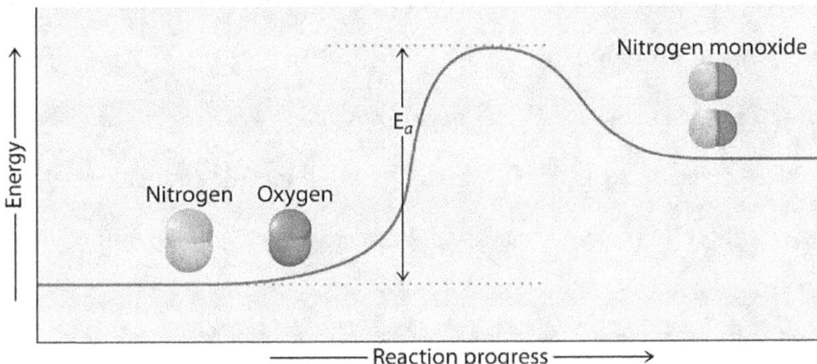

The reacting molecules require sufficient energy to overcome the energy barrier and transform into products.

Forward reaction:

> Activation energy E_a is the difference between the reactants' energy and the top of the reaction profile (i.e., transition state).

Reverse reaction:

> Activation energy, E_a, is the difference between the energy of the products and the top of the reaction profile (i.e., the transition state).

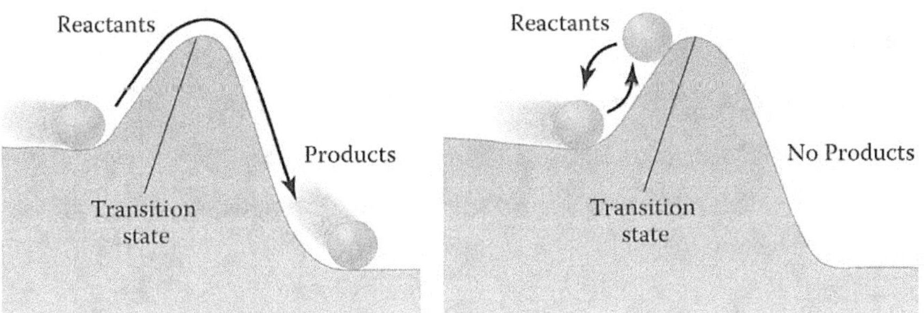

Exergonic reaction forms product   *Exergonic reaction without TS*

Unimolecular mechanisms

Unimolecular mechanisms involve *intermediates* (e.g., carbocation, radicals) during the reaction profile (two peaks on the reaction progress diagram).

Formation of an intermediate in first step is often the rate-determining step (i.e., slowest step).

The intermediate is consumed as the reaction forms products (via the second transition state).

Bimolecular mechanisms

Left image above is a completed bimolecular (E_2 or Sn_2) reaction.

Right image above shows a bimolecular reaction in which the activation energy (energy barrier) is not overcome: no transition state (TS) forms, and therefore, no products are generated.

Reaction mechanism influences rates

Most chemical reactions occur in several steps, known as a *reaction mechanism*.

Reaction mechanism is the sequence in which reactant molecules transform into products.

Each step in a multiple-step mechanism is an *elementary step* with an individual rate constant and rate law.

For example, the following reaction steps were experimentally observed:

$$(CH_3)_3CCl \ (aq) + OH^- \ (aq) \rightarrow (CH_3)_3COH \ (aq) + Cl^- \ (aq)$$

Step 1:

$$(CH_3)_3CCl \ (aq) \rightarrow (CH_3)_3C^+ \ (aq) + Cl^- \ (aq) \quad \text{(slow)}$$

$$r_1 = k[(CH_3)_3CCl]$$

Step 2:

$$(CH_3)_3C^+ \ (aq) + OH^- \ (aq) \rightarrow (CH_3)_3COH \ (aq) \quad \text{(fast)}$$

$$r_2 = k[(CH_3)_3C^+] \cdot [OH^-]$$

Rate-determining step

Rate-determining step is the slowest and has the most significant contribution to the reaction's overall rate.

In most cases, the overall reaction rate is almost equal to the rate-determining step's rate because the other steps are fast (almost instantaneous) and do not significantly affect the overall rate.

For example, the first step, the formation of the charged carbonium ion $(CH_3)_3C^+$, is slow compared to the second step, where the carbocation immediately reacts with the OH^- ion to form the neutral product $(CH_3)_3COH$.

$$(CH_3)_3CCl\,(aq) \rightarrow (CH_3)_3C^+\,(aq) + Cl^-\,(aq) \qquad \text{(slow – first step)}$$

$$(CH_3)_3C^+\,(aq) + OH^-\,(aq) \rightarrow (CH_3)_3COH\,(aq) \qquad \text{(fast – second step)}$$

Therefore, the first step is the rate-determining (slow) step.

Rate law of this (slow) step determines the overall rate:

$$r_1 = k[(CH_3)_3CCl]$$

There are reactions when the *rate-determining step* (or *slowest step*) is *not* the first step in the reaction mechanism.

In these examples, an intermediate may appear in the rate-determining step.

The intermediate must be substituted so that it is not in the rate law, requiring a detailed approach for determining the reaction mechanism's rate law.

Notes for active learning

Reaction Rate Depends on Temperature

Kinetic energy influences rate

Increasing temperature increases the average kinetic energy of colliding molecules; as a result, the molecules move faster.

Faster molecules are more likely to possess the energy necessary to overcome the activation energy and successfully collide and react.

Faster molecules (i.e., higher kinetic energy) collide at an increased frequency.

Therefore, according to collision theory, increasing the temperature increases the reaction rate.

However, some reactions might slow or stop as the temperature increases. This is common among reactions in biological systems because the enzyme proteins involved in the reaction may be damaged by higher temperatures, thereby reducing the enzyme's ability to participate in reactions.

Transition state as energy of activation

Transition state (or *activated complex*) is the energy peak (*y*-axis) of the energy profile when bonds begin to *break* (reactants) and other bonds begin to *form* (products).

Transition states form when the reactants have enough energy to achieve the required activation energy (the highest energy location on the reaction progress graph).

Molecules in the transition state *either* revert to reactants or form products.

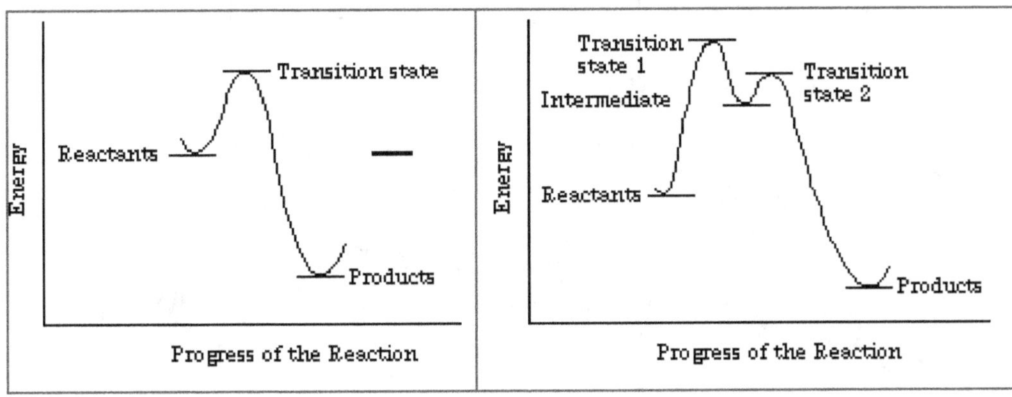

Bimolecular reactions   *Unimolecular reactions*

Note: Molecules in the transition state *cannot be isolated.*

Transition state bond making and breaking events

Transition states signify bond-breaking and bond-making events, as shown below.

Square brackets denote transition states [ ] and a *double dagger* ( ) around the fleeting (i.e., unable to be isolated) molecule.

Dashed lines symbolize temporary bonds that are simultaneously formed and broken, as indicated in brackets.

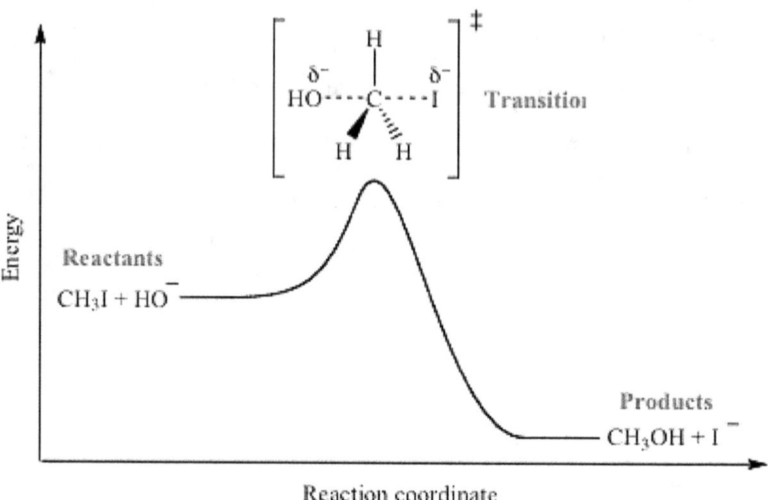

Energy profiles for activation energy E_{ac}

In a reaction, the products and reactants often have different energy levels.

Energy differential is released ($-\Delta H$) or absorbed ($+\Delta H$) from system as heat (enthalpy or ΔH).

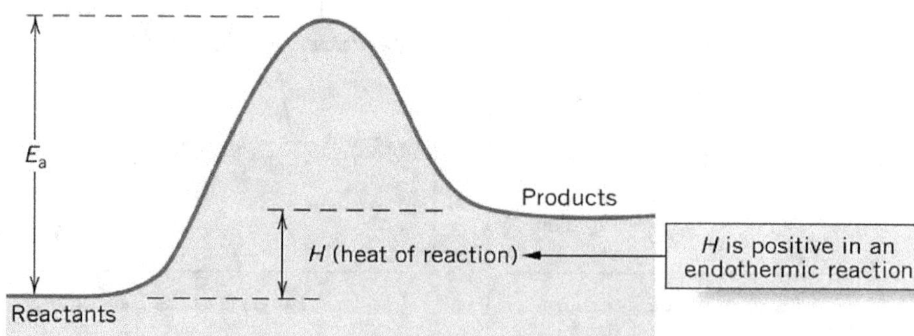

The graph's high point indicates the energy of the activation E_{ac} barrier during the reaction

Reactions are *endothermic* or *exothermic,* based on enthalpy.

Endothermic reactions,

> products have *higher potential energy* than reactants ($+\Delta H$), favoring *reverse reaction.*

Exothermic reactions,

> products have *lower potential energy* than reactants ($-\Delta H$), favoring *forward reaction.*

Forward reaction has a $-\Delta H$ (release energy) in the forward direction and a $+\Delta H$ in the reverse direction.

Enthalpy (ΔH)

Endothermic reactions have a potential energy of the products that is higher (i.e., less stable) than that of the reactants ($+\Delta H$).

> Heat is *required* as a reactant to drive the reaction forward.

> Heat is *absorbed* by the system from its surroundings to *break bonds.*

> An *energetically unfavorable* reaction occurs, and the temperature of the surroundings *decreases.*

Exothermic reactions have a potential energy of products that is *lower* (i.e., more stable) than for reactants ($-\Delta H$).

> Excess heat formed by the reaction is released by the system towards its surroundings.

> Energy is *released* when the system forms *bonds.*

> An *energetically favorable* reaction occurs, and the temperature of the surroundings *increases.*

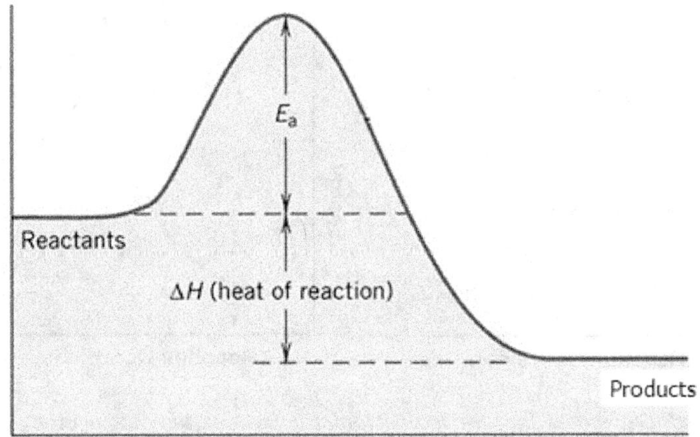

Exothermic reactions form products with lower energy than reactants

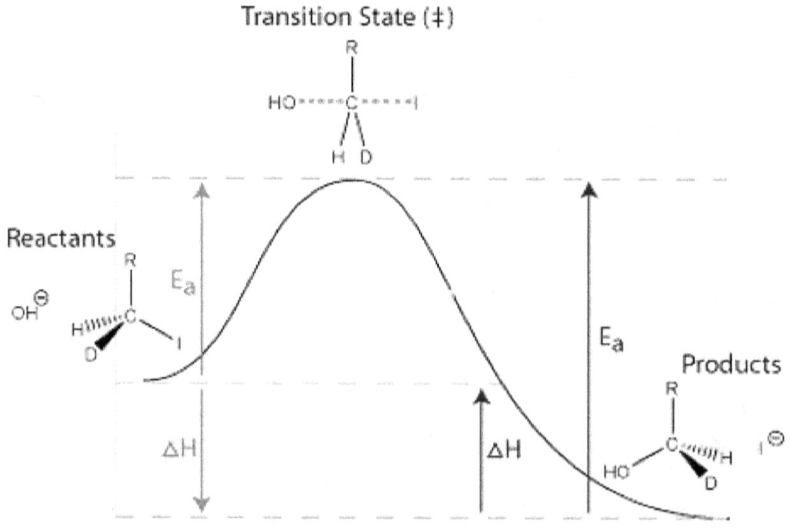

Energy profile diagrams summarize the reaction energies (ΔH and E_{ac})

Hammond postulate

Hammond's postulate (or *axiom*) states that *the transition state resembles either the reactants or the products, whichever is closer in energy.*

Exothermic reactions have a transition state (TS) *closer in energy to reactants than products.*

According to Hammond's postulate, the transition state *resembles the reactants.*

Endothermic reactions have a transition state that is *more energetic than the reactants but less energetic than the products.*

According to Hammond's postulate, the transition state *resembles the products.*

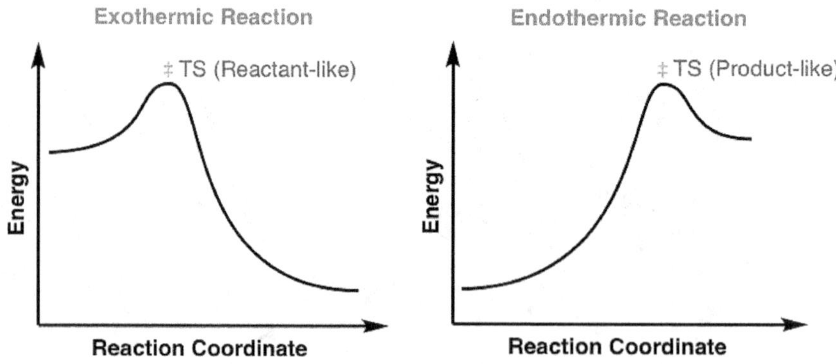

Hammond postulate states that the transition state resembles the species closer in energy

E_{ac} is lower (measured from the energy of reactants) than the peak of the energy barrier compared to the E_{ac} in the reverse reaction (measured from the energy of the products).

Arrhenius equation describes temperature-dependent rates

In 1889, Swedish Nobel laureate Svante Arrhenius (1859-1927) described the *temperature dependence* of reaction rates, whereby chemical reactions occur more rapidly at higher temperatures.

Arrhenius equation uses the relationship between rate constant (k), temperature (T), and activation energy (E_a):

$$k = Ae^{-E_{ac}/RT}$$

where k is the rate constant, R is the ideal gas constant ($R = 8.314$ J mol^{-1} K^{-1}), E_{ac} is the activation energy, T is the temperature in Kelvin, and A is the *Arrhenius frequency factor*

Arrhenius frequency factor is the pre-exponential factor (or collision factor) that represents the collision frequency.

For Arrhenius equation, ensure that the quantities' units are proper. T is in Kelvin, E_{ac} is in J/mol, and A has the same units as the rate constant k.

Point-slope form for the Arrhenius equation

Taking the natural logarithm of each side of the Arrhenius equation yields the point-slope form for the equation:

$$\ln k = -(E_a / RT) + \ln A$$

Straight-line slope is produced if data is graphed, where $\ln k$ is on y-axis, and $1 / T$ is on x-axis.

On the $\ln k$ *vs.* $1 / T$ plot, the line's slope is E_a / R, and the y-intercept corresponds to $\ln A$.

Determine slope of the line mathematically to calculate E_a, whereby the slope of a line is the change in y over the change in x (i.e., rise over run).

By extrapolation, the y-intercept and $\ln A$ collision factor are calculated.

Reaction rate constant at two temperatures is required to calculate the point-slope form.

Reaction rate constant relationship at *two temperatures* is expressed by:

$$\ln k_2 - \ln k_1 = \left(\ln A - \frac{E_a}{RT_2}\right)\left(\ln A - \frac{E_a}{RT_1}\right) = \frac{E_a}{R}\left(\frac{1}{T_1} - \frac{1}{T_2}\right)$$

The equation calculates the rate constant k of a reaction at two temperatures (in Kelvin).

y-intercept ($\ln A$) can be ignored because it is a constant and not part of the slope calculation.

Eliminate the $\ln A$ term by subtracting the expressions for the two $\ln$-k.

Simplified Arrhenius equation expresses activation energy

Since the slope embodies activation energy (E_{ac}), the simplified Arrhenius equation determines the E_{ac}.

Rearrange the expression to isolate activation energy.

$$E_a = \frac{R \ln \frac{k_2}{k_1}}{\frac{1}{T_1} - \frac{1}{T_2}}$$

For example, a reaction rate for a chemical process is investigated at two temperatures. The reaction rate at 25 °C is 1.55×10^{-4} s^{-1}. At 50 °C, the reaction rate is 3.88×10^{-4} s^{-1}. Based on this data, what is the energy of activation for the chemical process expressed in J/mol? (Use R = 8.314 J/mol·K)

Inspect units. The units for rates, s^{-1}, cancel. Since the activation energy is to be expressed in J/mol, the gas constant R is expressed as 8.314 J/mol·K. This choice of the gas constant, R, dictates the temperature units.

R constant has temperature units expressed in K; temperatures are converted to Kelvin:

$$T_1 = 25 \text{ °C} + 273 = 298 \text{ K}$$

$$T_2 = 50 \text{ °C} + 273 = 323 \text{ K}$$

$$E_a = \frac{R \ln \frac{k_2}{k_1}}{\frac{1}{T_1} - \frac{1}{T_2}}$$

Substitute the rate constants and temperatures into the Arrhenius equation:

$$E_{ac} = \frac{8.314 \text{ J/mol·K} \times \ln(3.88 \times 10^{-4} \text{ s}^{-1} / 1.55 \times 10^{-4} \text{ s}^{-1})}{(1/298 \text{ K}) - (1/323 \text{ K})}$$

$$E_{ac} = \frac{8.314 \text{ J/mol·K} \times \ln(2.503)}{(2.6 \times 10^{-4} \text{ K})}$$

$$E_{ac} = 2.93 \times 10^4 \text{ J/mol}$$

Van't Hoff equation for K_{eq}

Van't Hoff equation relates the change in the equilibrium constant (K_{eq}) of a chemical reaction to the change in temperature.

Van't Hoff equation calculates the *equilibrium constant K* (uppercase K for equilibrium; lowercase k denotes rate) of a reaction at *two Kelvin temperatures*.

Van't Hoff equation modifies the Arrhenius equation by replacing E_{ac} with ΔH.

This modified Arrhenius equation is the Van't Hoff equation:

$$\ln\left(\frac{K_{T_2}}{K_{T_1}}\right) = \frac{\Delta H^{\circ}}{R}\left(\frac{1}{T_1} - \frac{1}{T_2}\right)$$

The above equation is *not* the Arrhenius equation.

Notes for active learning

Kinetic *vs.* Thermodynamic Control of Reactions

Reaction selectivity

Several chemical reactions have distinct products, and *reaction conditions* affect selectivity.

Thermodynamics explains how ΔG and other variables affect whether a reaction occurs spontaneously.

Kinetics describe how fast a reaction occurs (based on the activation energy), and kinetics has no impact on the spontaneity or thermodynamics of the reaction.

> *Kinetic product*s require lower activation energy and are formed preferentially (according to the transition state being of lower energy) at lower temperatures.

> *Thermodynamic products* have a more negative ΔG (favorable) and are formed preferentially (according to higher energy transition state) at higher temperatures.

Energy profile diagrams display a reaction under kinetic *vs.* thermodynamic control.

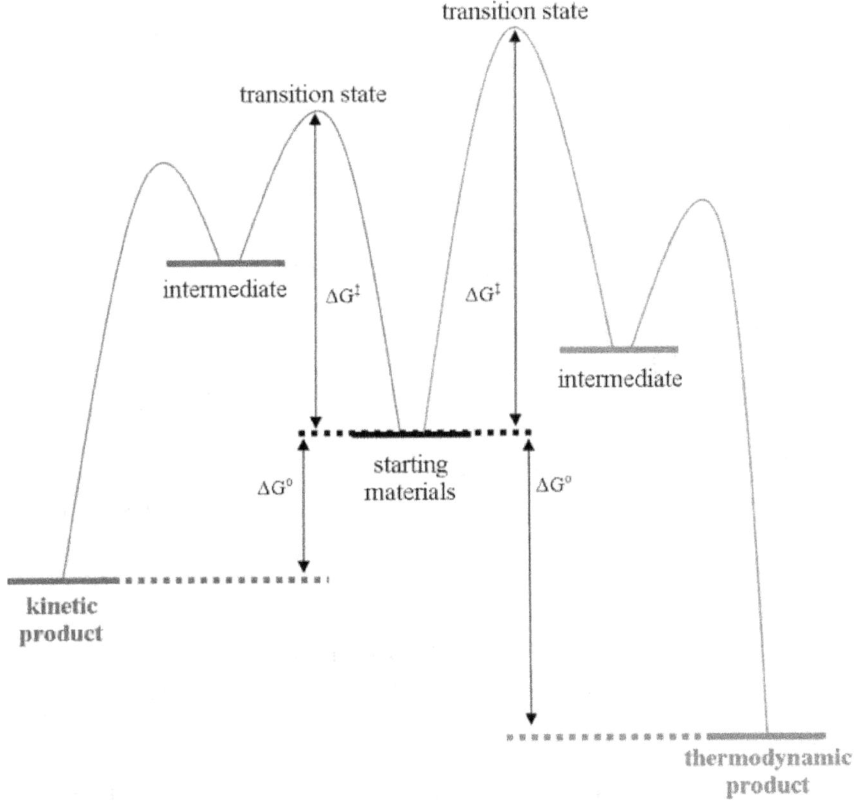

Kinetic vs. thermodynamic control for products depends on energy of transition state

Reactants (i.e., starting material) are in the center of the graph above.

On the left is an intermediate with lower energy than the intermediate on the right.

The intermediate with lower energy forms *faster* (I.e., kinetic control).

However, the final product from this lower energy intermediate yields a *higher energy product* than the more stable product on the right (i.e., thermodynamic product).

Product on the right on the graph above is the thermodynamic product (overall more stable than the kinetic product). However, it requires a higher (more unstable) transition state before forming a product.

> *Kinetic* products form *fast* because of a *lower energy* transition state.

> *Thermodynamic* products form *slowly* because of a *higher energy* transition state.

Reaction mechanism drives product formation

However, the ΔG, the transition states, and the energies of the intermediates and transition states are different.

The product is different, depending on whether the reaction (rxn) is under kinetic control (top) or thermodynamic control (bottom).

When strong acids are added to certain conjugated dienes (e.g., butadiene), two products form: the 1,2- product and the 1,4-product.

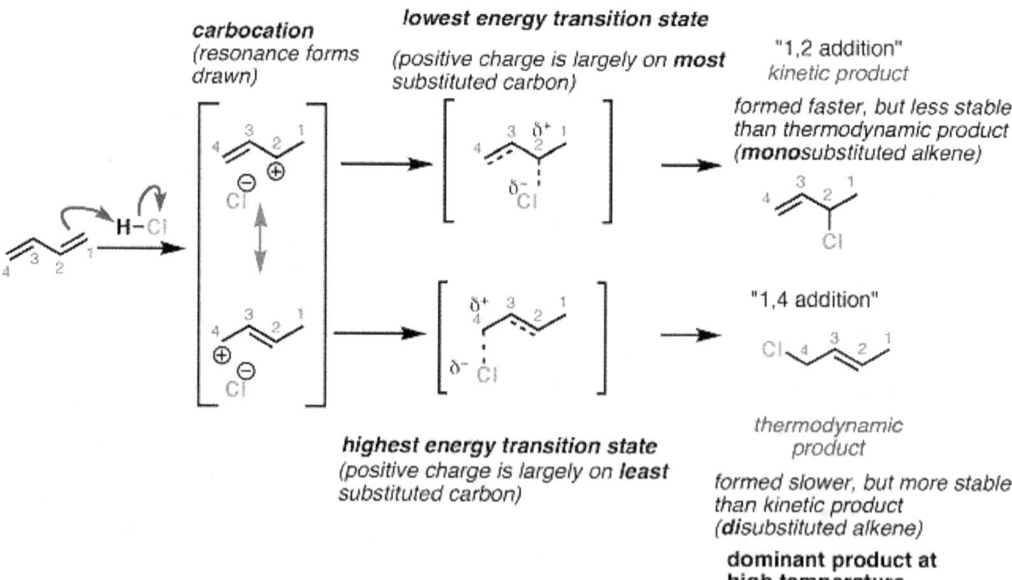

1,2 vs. 1,4 addition of conjugated dienes form kinetic (less stable) and thermodynamic (more stable) products

In the top transition state, the carbocation's positive charge is localized on the most substituted (2°) carbon, and the transition state has lower energy.

Lower potential energy transition state indicates that the reaction has lower activation energy.

Kinetic (1,2) product

1,2 *kinetic product* is preferentially formed at *lower temperatures*.

The bottom transition state is higher in energy because the carbocation's positive charge is localized on the less substituted (1°) carbon.

Therefore, this transition state has higher energy. This intermediate indicates that the bottom reaction has a higher activation energy.

Thermodynamic (1,4) product

1,4 *thermodynamic product* is preferentially formed at higher temperatures, which are necessary to proceed with the less stable (2°) carbon intermediate.

1,4-product is more stable than the 1,2-product because it is an internal alkene instead of a terminal alkene.

1,4-addition produces the thermodynamically stable product (i.e., lower potential energy) but proceeds more slowly through the transition state, which is of higher energy (less substituted carbocation).

1,4-addition product is the major product formed at higher temperatures, which are necessary to proceed through the less stable transition state.

Terminal *vs.* internal alkene

The energy profile diagram for an E_1 *vs.* E_2 reaction is shown:

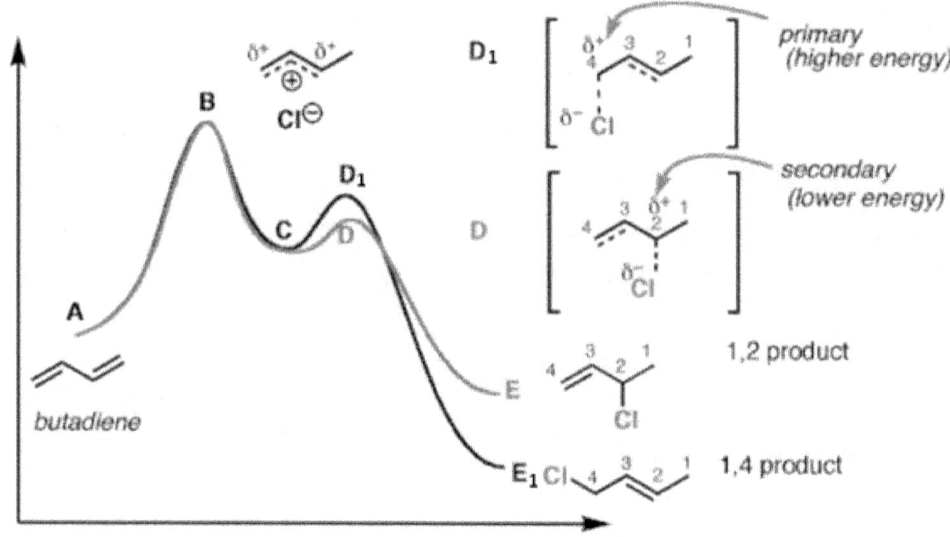

The energy of HCl addition to 1,2-addition vs. 1,4-addition to butadiene

The height of transition states D and D_1 determines reaction rate once the carbocation C forms.

Lower energy transition states (D *vs.* D_1) proceed faster to products.

Therefore, E (terminal alkene) forms faster from the secondary carbocation intermediate C since the energy of transition is less for D (secondary carbocation) than D_1 (primary carbocation).

Energy of E (terminal alkene) and E_1 (internal alkene) is related to greater stability for the 1,4-addition compared to the 1,2-addition.

E_1 has a more substituted (internal) double bond than E (terminal alkene), which is more stable.

E_1 represents the thermodynamic (more stable) product that is *slower* to form (higher energy transition state) but is *more stable* than the kinetic product E, which forms faster (lower energy transition state) but is less stable than E_1.

Catalysts and Enzymes

Catalysts lower energy of activation

Catalysts are substances in a chemical reaction that increase the reaction rate by lowering the activation energy (often through an alternative pathway) needed to proceed.

Catalyst participate in chemical reactions without being chemically altered and used repeatedly.

For a reaction mechanism separated into elementary steps, a catalyst appears at the start and reappears at the end of the reaction. A catalyst is neither a reactant nor a product in the reaction.

Acids, bases, and metal ions often act as catalysts.

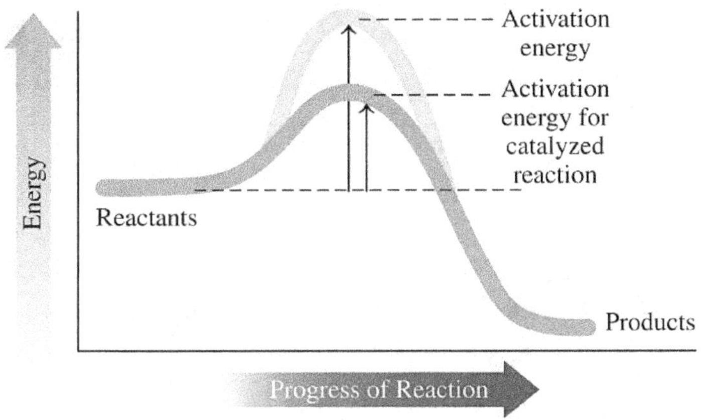

Homogeneous *vs.* heterogeneous catalysts

There are homogeneous and heterogeneous catalysts.

Homogeneous catalysts are in the same phase as the reactants.

Heterogeneous catalysts are in a different phase from the reactants. These catalysts often immobilize reactants close to each other, increasing the likelihood of a collision and, consequently, the corresponding reaction rate.

An example of a heterogeneous catalyst is a dissolved acid that catalyzes the hydrolysis of esters in an aqueous solution.

Most heterogeneous catalysts are solids, and the reactants are often gases or liquids.

For example, iron catalyzes the synthesis of NH_3 from gaseous N_2 and H_2.

Modern cars have *catalytic converters* in their exhaust systems, which serve as heterogeneous catalysts, providing a solid surface for exhaust molecules to bind and react. Catalytic converters reduce pollutants such as carbon monoxide (CO), nitrogen oxides (NOx), and hydrocarbons (e.g., octane, C_8H_{18}).

Catalysts usually consist of solid particles, including platinum (Pt) and palladium (Pd).

For example, the reaction:

$$2\,NO\,(g) \rightarrow N_2\,(g) + O_2\,(g)$$

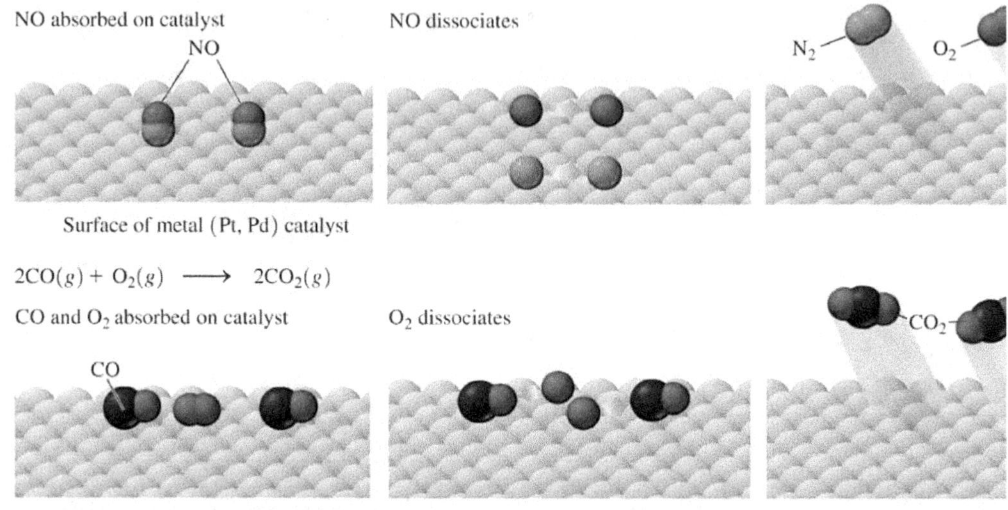

Biological enzymes

Enzymes (e.g., proteins) are *biological catalysts* that increase the reaction rates of biochemical reactions.

Biological enzymes are more efficient than chemical catalysts; enzymes can increase a reaction rate by a factor of more than ten million (1×10^7).

Enzymes are highly specific; they form enzyme-substrate complexes with reactants.

Substrate is a molecule on which the enzymes act.

For example, when hydrogen peroxide (H_2O_2) is applied to a wound, oxygen (O_2) gas is produced by decomposing hydrogen peroxide by the catalase enzyme in the blood.

Decomposition of hydrogen peroxide by catalase:

$$2\,H_2O_2\,(l) \xrightarrow{catalase} 2\,H_2O\,(l) + O_2\,(g)$$

Catalysts are written above or below the arrow in the reaction.

Catalyzed *vs.* non-catalyzed reactions

Changing variables (i.e., reactant concentration and temperature) can increase the reaction rate.

Catalyst has a similar effect on the reaction rate but uses a different method.

Factor	Result
Increasing reactant concentration	More collisions
Increasing temperature	More collisions exceed the energy of activation
Adding a catalyst	Lowers the energy of activation

Notes for active learning

Equilibrium in Reversible Reactions

Reversible reactions

Some chemical reactions are reversible; they proceed in either direction, resulting in an equilibrium mixture of reactants and products.

If a chemical bond forms, it can be broken with sufficient energy input.

Reversible chemical reactions are indicated by a *double-headed arrow* or two arrows facing opposite directions.

$$A + B \leftrightarrow AB$$

$$A + B \leftrightarrows AB$$

Forward reaction occurs when A and B form AB.

When AB (product) accumulates, the reverse reaction occurs from the decomposition of AB, and A and B (reactants) are reformed.

When the forward and the reverse reactions have the *same rate*, the reaction is in chemical *equilibrium*.

Equilibrium

Equilibrium does not mean equal amounts of reactants and products but is the point where the number of reactants and products *remains constant* because the *forward and reverse reactions proceed at equal rates*.

For example, consider the following reaction:

$$2\,SO_2\,(g) + O_2\,(g) \;\rightleftarrows\; 2\,SO_3\,(g)$$

Reaction proceeds forward with SO_2 and O_2, and SO_3 products form until equilibrium (i.e., the equal rate for the forward and reverse reaction).

From SO_3 products, the reaction proceeds in the reverse direction, forming SO_2 and O_2 until equilibrium is reached.

As shown, the equilibrium concentrations of SO_2, O_2, and SO_3 are the same for forward and reverse reactions.

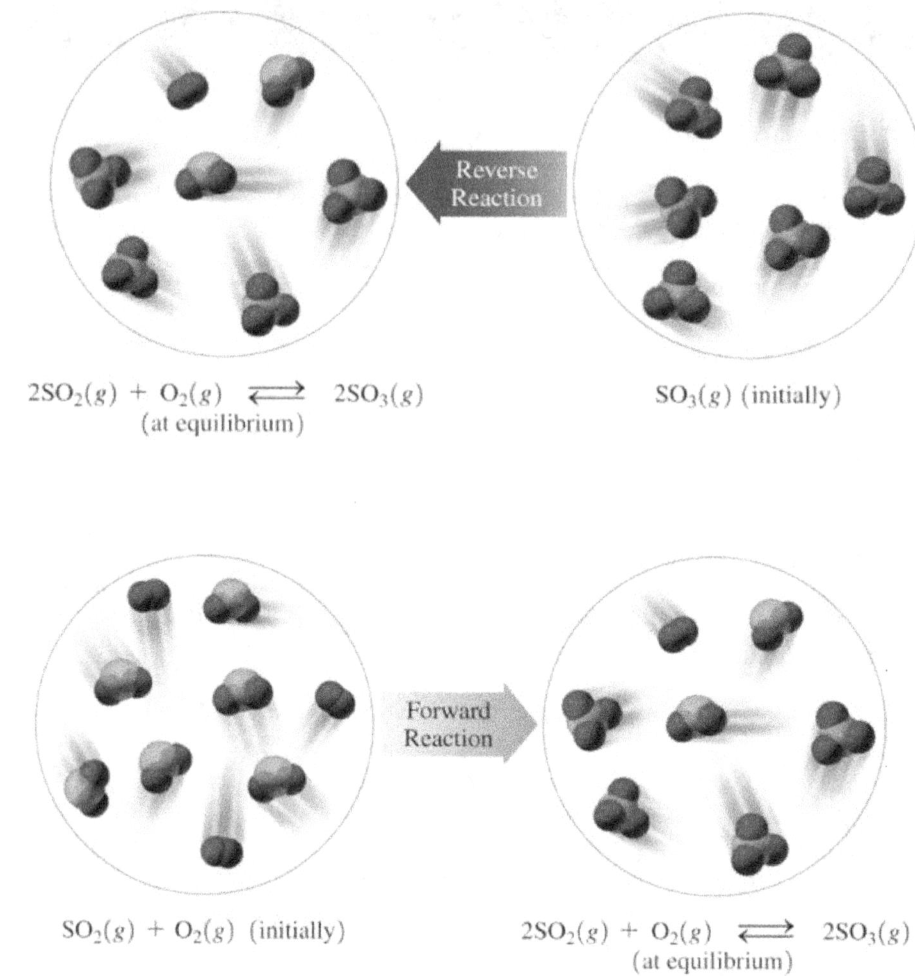

$2SO_2(g) + O_2(g) \rightleftharpoons 2SO_3(g)$
(at equilibrium)

$SO_3(g)$ (initially)

$SO_2(g) + O_2(g)$ (initially)

$2SO_2(g) + O_2(g) \rightleftharpoons 2SO_3(g)$
(at equilibrium)

Experimental evidence for equilibrium

For example, consider the reversible reaction that occurs during the evaporation and condensation of water. Suppose water is placed over a gas-filled chamber, separated by a moving piston as diagrammed.

Work *on* or *by* the system is observed by measuring the gas' compression or expansion.

As water evaporates, the water's mass decreases, and the gas expands.

When water vapor condenses, it increases the weight of water and *compresses the gas*.

Once the system reaches equilibrium, the liquid-to-vapor ratio remains constant. The gas neither compresses nor expands.

Therefore, *no work* is done *on* or *by* the system at equilibrium.

Water molecules constantly transform between liquid and gas phases, but the evaporation rate matches the condensation rate, so there is no noticeable change in composition.

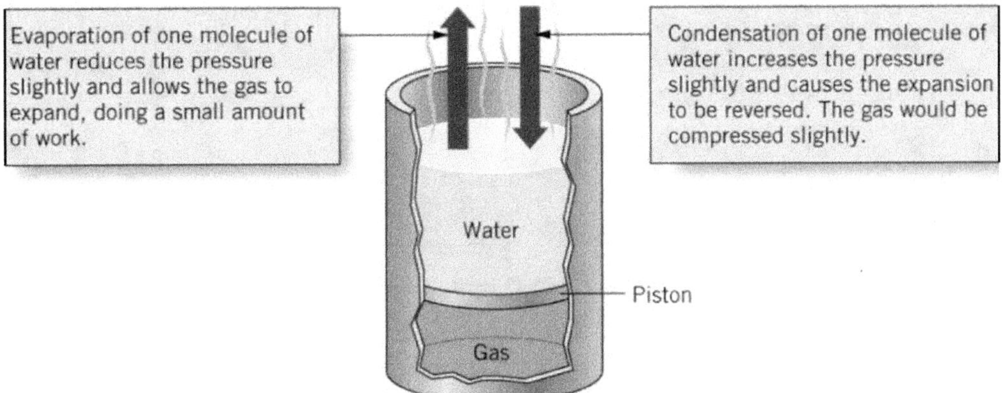

Piston system detects the equilibrium for water

Notes for active leaning

Mass Action

Law of Mass Action

Law of Mass Action states that the *reaction rate depends on the concentration of the participating substances*.

$$aA + bB \rightleftarrows cC + dD$$

$$\text{rate}_{\text{rxn}} = \left(-\frac{1}{a}\right)\frac{\Delta A}{\Delta t} = \left(-\frac{1}{b}\right)\frac{\Delta B}{\Delta t} = \left(+\frac{1}{c}\right)\frac{\Delta C}{\Delta t} = \left(+\frac{1}{d}\right)\frac{\Delta D}{\Delta t}$$

Deriving the equilibrium constant

Law of Mass Action *derives the equilibrium constant* by setting the forward reaction rate equal to the reverse reaction rate (i.e., equilibrium).

For the reaction:

$$aA + bB \rightleftarrows cC + dD$$

Relationships exist:

$$r_{\text{forward}} = r_{\text{reverse}}$$

$$k_{\text{forward}} \cdot [A]^a \cdot [B]^b = k_{\text{reverse}} \cdot [C]^c \cdot [D]^d$$

$$K_c = k_{\text{forward}} / k_{\text{reverse}} = [C]^c \cdot [D]^d / [A]^a \cdot [B]^b$$

Phases for Law of Mass Action

Law of Mass Action is based on the equilibrium constant K_c, which is the ratio of the concentration of products to concentration of reactants (i.e., [products] / [reactants]).

Mass action expression of equilibrium *only* includes equilibrium components in the *same phase*.

Pure solids and liquids do *not* appear in the mass action expression.

The absence of pure solids and liquids from the equation is:

$$CaCO_3\ (s) \rightleftarrows CaO\ (s) + CO_2\ (g)$$

$$K_c = [CO_2]$$

For example, consider the following equation:

$$2\ HCl\ (aq) + CaCO_3\ (s) \rightleftarrows CaCl_2\ (aq) + CO_2\ (g) + 2\ H_2O\ (l)$$

Net ionic equation for the above reaction is:

$$2\,H^+\,(aq) + CaCO_3\,(s) \rightleftharpoons Ca^{2+}\,(aq) + CO_2\,(g) + 2\,H_2O\,(l)$$

Cl^- ions were present on each side; therefore, they are spectator ions and *removed* from the net equation.

Mass Action equilibrium constants

Mass action equilibrium constant has two forms.

K_{c1} is expressed as concentration of the aqueous species.

$$K_{c1} = [Ca^{2+}] / [H^+]^2$$

K_{c2} is expressed as concentration of the gaseous species.

$$K_{c2} = [CO_2]$$

The K_{c1} and K_{c2} constants have different values.

Commonly reported K values are based on the most accessible phase that can be measured experimentally.

Do not mix phases, and do not include solids in either expression. When solving mass action problems, check the phases of the molecules.

Equilibrium Constants

Equilibrium constant K_c

Equilibrium constant K_c is the ratio of the molar concentrations of products divided by reactants at equilibrium.

K_c is abbreviated as K_{eq}, a general term for equilibrium constants that represent concentrations or partial pressures.

For the reaction $a\text{A} + b\text{B} \rightleftharpoons c\text{C} + d\text{D}$, the equilibrium constant is:

$$K_c = \frac{[\text{C}]^c[\text{D}]^d}{[\text{A}]^a[\text{B}]^b} = \frac{[\text{products}]}{[\text{reactants}]}$$

The equilibrium constant K_c is a temperature-specific constant, and it retains the same value regardless of the molar concentrations, provided the temperature remains constant.

The units of K_c depend on the reaction; K_c itself has no units.

Calculating K_c

Calculating the K_c value:

1. State the given and needed qualities.

2. Write the K_c expression for the equilibrium.

3. Substitute equilibrium (molar) concentrations and calculate K_c.

For example, what is the value of K_c at 443 °C for the equilibrium concentrations:

$$\text{H}_2\,(g) + \text{I}_2\,(g) \rightleftharpoons 2\,\text{HI}\,(g)$$

$[\text{H}_2] = 1.2$ mol/L

$[\text{I}_2] = 1.2$ mol/L

$[\text{HI}] = 0.35$ mol/L

Step 1: State the given and needed quantities

Given values		Needed quantities
Reactants	*Products*	
$[\text{H}_2] = 1.2$ mol/L	$[\text{HI}] = 0.35$ mol/L	
$[\text{I}_2] = 1.2$ mol/L		K_c

Step 2: Write the K_c expression for the equilibrium

$$K_c = [\text{HI}]^2 / [\text{H}_2]\cdot[\text{I}_2]$$

Step 3: Substitute equilibrium (molar) concentrations and calculate K_c

$$K_c = [0.35]^2 / [1.2]\cdot[1.2]$$

$$K_c = 8.5 \times 10^{-2}$$

K_c magnitude indicates reaction direction

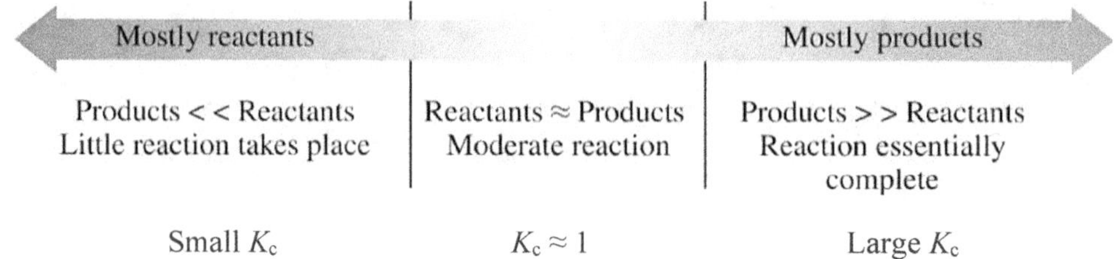

K_c values depend on whether the equilibrium has a greater quantity of products or reactants.

However, the equilibrium constant's magnitude does not affect how fast equilibrium is reached.

Reactions with a large K_c have *more products* created from the forward reaction, and products predominate at equilibrium.

Reactions with small K_c have *reactants* that predominate at equilibrium.

For example, the equilibrium for the reaction of SO_2 and O_2 has a large K_c.

$$2\,SO_2\,(g) + O_2\,(g) \leftrightarrows 2\,SO_3\,(g)$$

At equilibrium, the reaction contains mostly *products*.

$$K_c = [\text{SO}_3]^2 / [\text{SO}_2]^2\cdot[\text{O}_2]$$

$$K_c = [\text{many products}] / [\text{few reactants}]$$

$$K_c = 3.4 \times 10^2$$

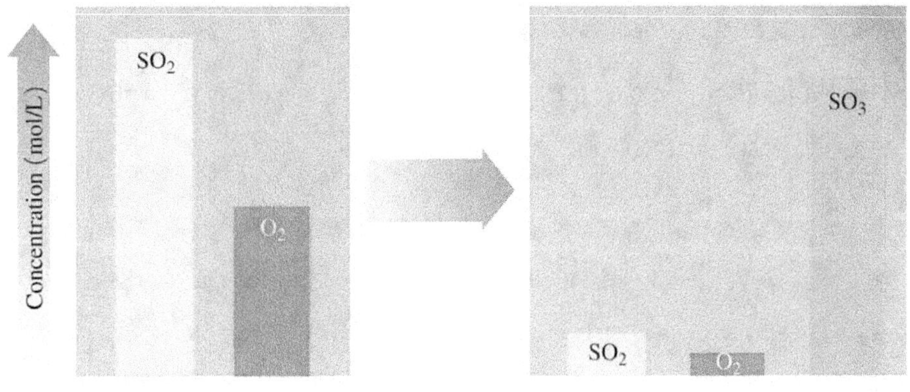

For example,

$$2\ SO_2\ (g) + O_2\ (g) \rightleftarrows 2\ SO_3\ (g)$$

Reactions with a small K_c have an equilibrium mixture with a *low concentration of products* and a *high concentration of reactants*.

Equilibrium constant for the reaction of N_2 and O_2 has a small K_c.

$$N_2\ (g) + O_2\ (g) \rightleftarrows 2\ NO\ (g)$$

At equilibrium, the reaction mixture contains mostly *reactants*.

$$K_c = [NO] / [N_2] \cdot [O_2]$$

$$K_c = [\text{few products}] / [\text{mostly reactants}]$$

$$K_c = 2 \times 10^{-9}$$

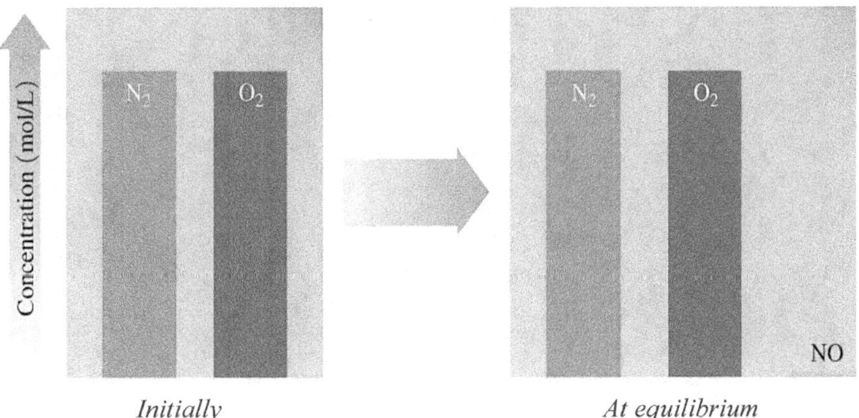

Homogeneous *vs.* heterogeneous equilibrium

Homogeneous equilibrium is a reaction with products and reactants in same physical state:

$2 SO_2 (g) + O_2 (g) \rightleftarrows 2 SO_3 (g)$

$K_c = [SO_3]^2 / [SO_2]^2 \cdot [O_2]$

Heterogeneous equilibrium is a reaction with substances in different physical states:

$C (s) + H_2O (g) \rightleftarrows CO (g) + H_2 (g)$

$K_c = [CO] \cdot [H_2] / [H_2O]$

Concentrations of liquids and solids do not change and are omitted from equilibrium constant K_c expression.

Note: Omission of liquids and solids is like a mass-action equilibrium.

Reaction quotient Q_c

Reaction quotient Q_c is calculated using the same method as Kc, but Qc can be calculated during the reaction.

Q_c is a "snapshot" of the system, representing the ratio between products and reactants at a given time.

Comparing Q_c and K_c determines the direction of the reaction:

If $Q_c > K_c$, the reaction creates more reactants

If $Q_c < K_c$, the reaction creates more products

If $Q_c = K_c$, the reaction is at equilibrium, $\Delta G = 0$

Equilibrium Variables

Le Châtelier's Principle for equilibrium

Le Châtelier's Principle states that when a reversible reaction at equilibrium is perturbed by a change in a variable (e.g., concentration, pressure, volume, or temperature), the equilibrium shifts to counteract the effect of the change.

For the equilibrium between colorless N_2O_4 and brown NO_2:

$$N_2O_4 \, (g) \rightleftarrows 2 \, NO_2 \, (g)$$

Effect of concentration on equilibrium

If N_2O_4 *reactants increase*, the reaction shifts to the right to produce more NO_2.

If NO_2 *products increase*, the reaction shifts to the left, producing more N_2O_4.

Effect of pressure on equilibrium

Increasing pressure shifts reaction to side with fewer gas molecules in a gaseous equilibrium.

For the reaction:

$$N_2O_4 \, (g) \rightleftarrows 2 \, NO_2 \, (g)$$

Increasing pressure shifts the reaction to the left, producing N_2O_4.

Changing pressure does *not* shift equilibrium for equal moles of gases as reactants and products.

Effect of decreasing volume on equilibrium

A change in the volume of a gas mixture at equilibrium changes the concentration of gases in the mixture.

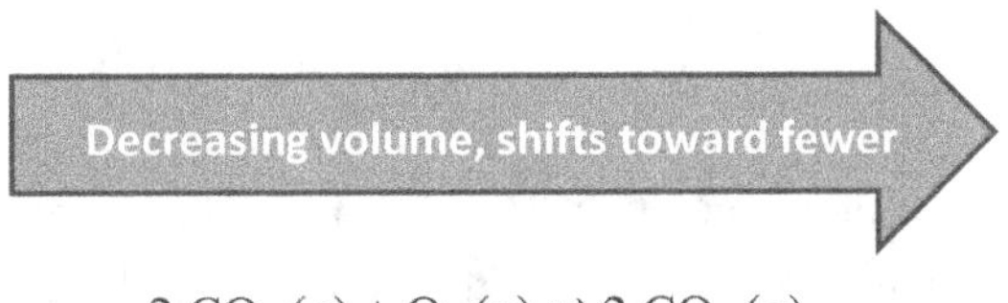

$$2 \, CO_2 \, (g) \mid O_2 \, (g) \rightleftarrows 2 \, CO_2 \, (g)$$

Decreasing the volume increases the concentration of the gases as the system shifts toward the smaller number of moles due to the decrease in volume.

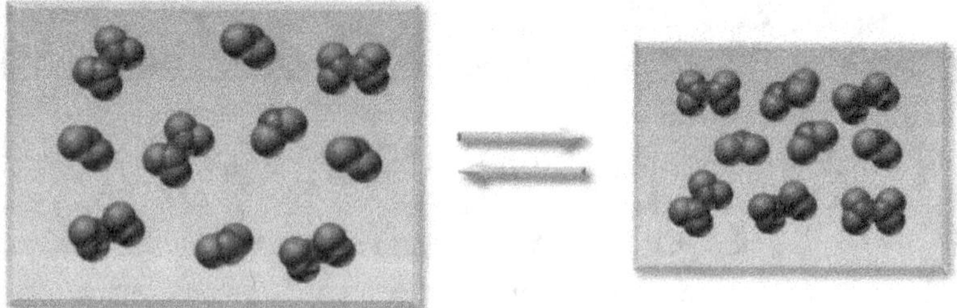

<table>
<tr><td>*1.00 L at equilibrium*</td><td>*0.750 L at equilibrium*</td></tr>
</table>

Effect of increasing volume on equilibrium

A change in the volume of a gas mixture changes its concentration.

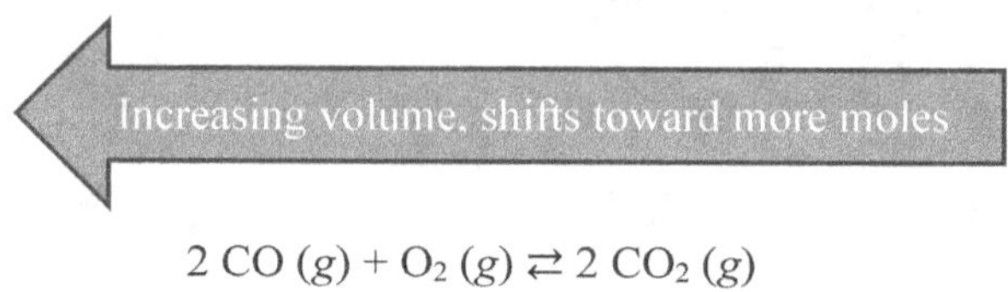

$$2\ CO\ (g) + O_2\ (g) \rightleftarrows 2\ CO_2\ (g)$$

Increasing the volume decreases gas concentration as the system shifts toward a larger number of moles to compensate.

Temperature effect on equilibrium depends on the ΔH (i.e., enthalpy) of a reaction (i.e., endothermic or exothermic).

Endothermic reaction equilibrium and temperature

Decreasing the temperature of an endothermic reaction ($+\Delta H$ with heat as a reactant) causes the system to shift the reaction toward heat, thereby increasing the heat in the system.

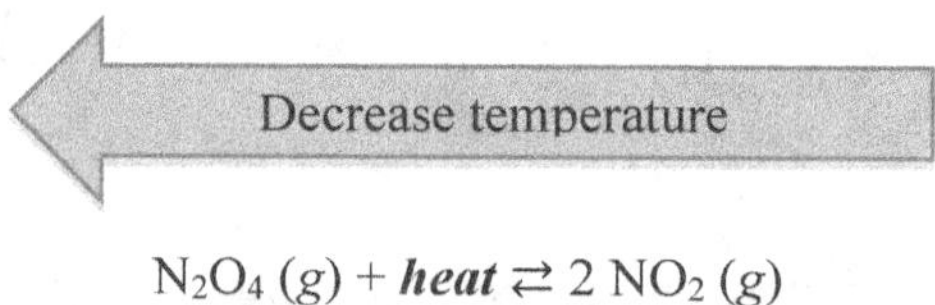

$$N_2O_4\ (g) + \textbf{heat} \rightleftarrows 2\ NO_2\ (g)$$

Increasing the temperature of an endothermic reaction ($+\Delta H$ with heat as a reactant) causes the system to shift the reaction toward the products that consume the heat.

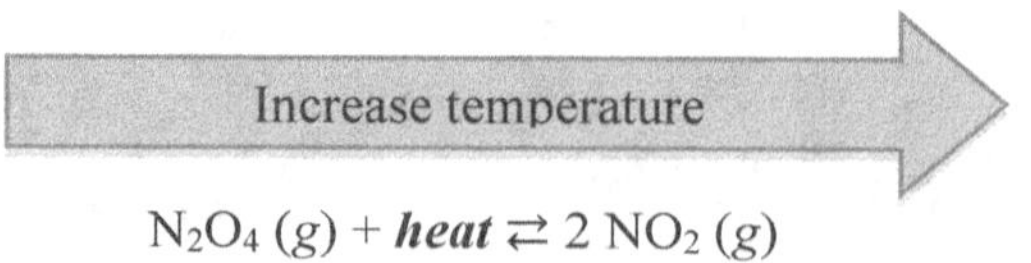

$$N_2O_4\ (g) + \textbf{heat} \rightleftarrows 2\ NO_2\ (g)$$

Exothermic reaction equilibrium and temperature

Decreasing the temperature of an exothermic reaction ($-\Delta H$ with heat as a product) causes the system to shift the reaction toward the products, thereby increasing the heat in the system.

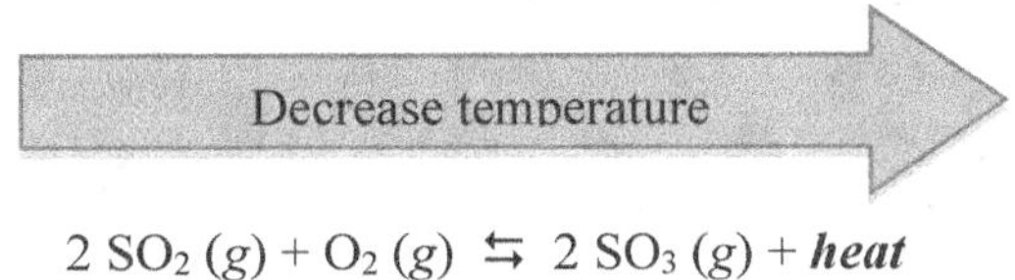

$$2\ SO_2\ (g) + O_2\ (g) \rightleftarrows 2\ SO_3\ (g) + \textbf{\textit{heat}}$$

Increasing the temperature of an exothermic reaction ($-\Delta H$ with heat as a product) causes the system to shift the reaction toward removing heat; it shifts the reaction toward the reactants, thereby decreasing the heat in the system.

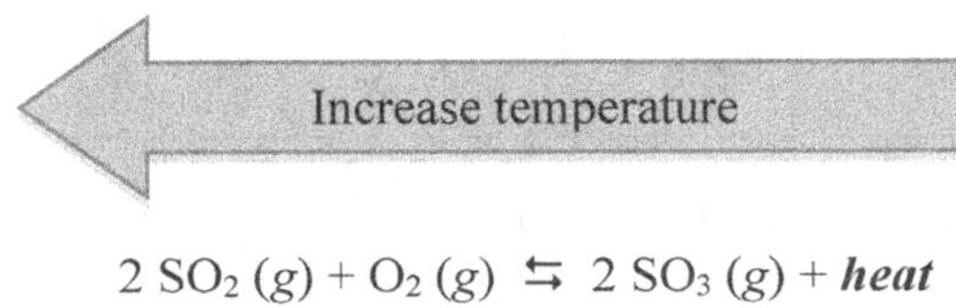

$$2\ SO_2\ (g) + O_2\ (g) \rightleftarrows 2\ SO_3\ (g) + \textbf{\textit{heat}}$$

Variables affecting equilibrium

Variable	Stress	Equilibrium shifts toward
Concentration	Add a reactant	Products (forward reaction)
	Remove a reactant	Reactants (reverse reaction)
	Add a product	Reactants (reverse reaction)
	Remove a product	Products (forward reaction)
Volume (container)	Decrease volume	Side with lower total coefficients (*aq* and *g*)*
	Increase volume	Side with higher total coefficients (*aq* and *g*)
Pressure	Decrease pressure	Side with higher total coefficients (*aq* and *g*)
	Increase pressure	Side with lower total coefficients (*aq* and *g*)
Temperature	**Endothermic Rxn**	
	Raise Temperature	Products (forward reaction to *remove heat*)
	Lower Temperature	Reactants (reverse reaction to *add heat*)
	Exothermic Rxn	
	Raise Temperature	Reactants (reverse reaction to *add heat*)
	Lower Temperature	Products (forward reaction to *remove heat*)
Catalyst	Increases rates equally	No effect

* Pure solids and pure liquids are not included in the equilibrium expression.

Biological equilibrium

Equilibrium is a crucial concept in various systems, including biological ones.

For example, O_2 transport uses an equilibrium between hemoglobin (Hb), oxygen, and oxyhemoglobin (HbO_2).

$$Hb\ (aq) + O_2\ (g) \rightleftarrows HbO_2\ (aq)$$

$$K_c = [\text{products}] / [\text{reactants}]$$

$$K_c = [HbO_2] / [Hb]\cdot[O_2]$$

If there is a high concentration of O_2 (reactant) in the lungs' alveoli, the reaction shifts to the right to make more oxyhemoglobin (product).

When the concentration of O_2 ($[O_2]$) is low in tissues, the reverse reaction releases O_2 from oxyhemoglobin.

At typical atmospheric pressure, oxygen diffuses into the blood because the partial pressure of oxygen in the alveoli is higher than that in the blood.

At altitudes above 8,000 ft, a decrease in atmospheric pressure results in the lower partial pressure of O_2.

Hypoxia may occur at high altitudes where the $[O_2]$ is low. At an altitude of 18,000 ft., a person obtains 29% less oxygen and may experience hypoxia.

According to Le Châtelier's Principle, a decrease in O_2 shifts the equilibrium in the direction of the reactants and depletes the concentration of HbO_2, which causes hypoxia.

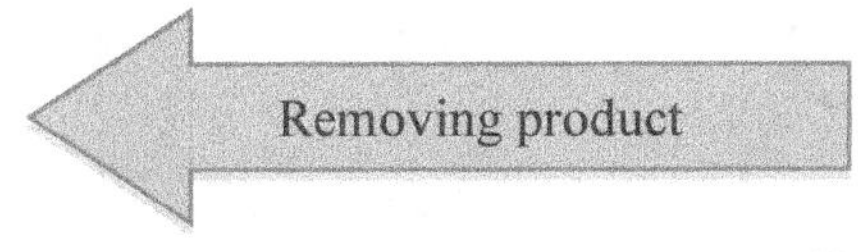

$$Hb\ (aq) + O_2\ (g) \rightleftarrows\ HbO_2\ (aq)$$

Another reaction that applies Le Châtelier's Principle follows.

For example, indicate the shift in equilibrium caused by each change:

$$2\ NO_2\ (g) + heat \rightleftarrows 2\ NO\ (g) + O_2\ (g)$$

Which change causes the equilibrium to shift toward reactants or products (*answers below*)?

A. adding NO
B. lowering the temperature
C. removing O_2
D. increasing the volume
E. removing NO

Answers:

A: reactants	C: products	
B: reactants	D: products	E: products

Equilibrium Constant and ΔG° Relationship

Direction and extent of reaction

Gibbs free energy ($\Delta G°$) is an expression for a system's free energy.

Equations relate Gibbs free energy to equilibrium constant:

$$\Delta G° = -RT \ln K$$

Taking the antilog (e^x) of each side gives:

$$K = e^{-\Delta G° / RT}$$

ΔG and K imply the direction and extent of a reaction, *not* the rate.

ΔG at equilibrium

If $\Delta G° = 0$, the reaction is at equilibrium; no net change to the system or surroundings.

$$\Delta G° = 0$$

Reaction appears spontaneous in either direction if a slight change is made to the system.

System remains at equilibrium if no heat is added or removed.

Phases exist indefinitely.

ΔG affects equilibrium

ΔG affects the position of equilibrium in the following ways:

$$\Delta G° > 0 \text{ (positive)}$$

If $\Delta G° > 0$, equilibrium is closer to reactants; reverse reaction (towards reactants) is spontaneous.

$$\Delta G° < 0 \text{ (negative)}$$

If $\Delta G° < 0$, equilibrium is closer to products; forward reaction (towards products) is spontaneous.

For example, consider freezing water at 0 °C:

$$H_2O\ (l) \rightarrow H_2O\ (s)$$

Below 0 °C:

$$\Delta G < 0 \text{ and freezing is spontaneous}$$

Above 0 °C:

$$\Delta G > 0 \text{ and freezing is nonspontaneous}$$

Energy profile for spontaneous reactions

Energy profile displays a reaction with a $+\Delta G$ (i.e., nonspontaneous).

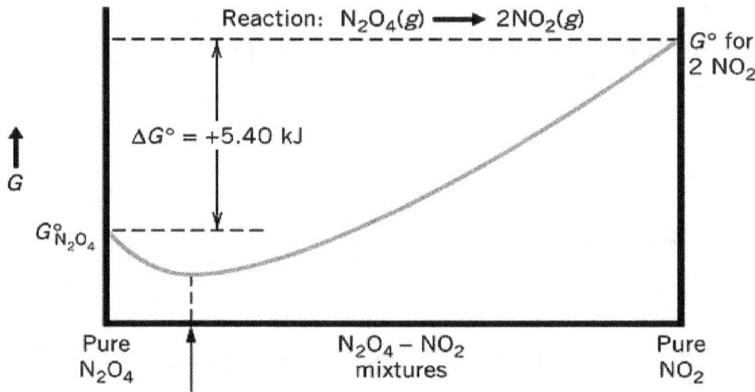

Energy profile for nonspontaneous reactions

The following energy profile represents a reaction with $-\Delta G$ (i.e., spontaneous).

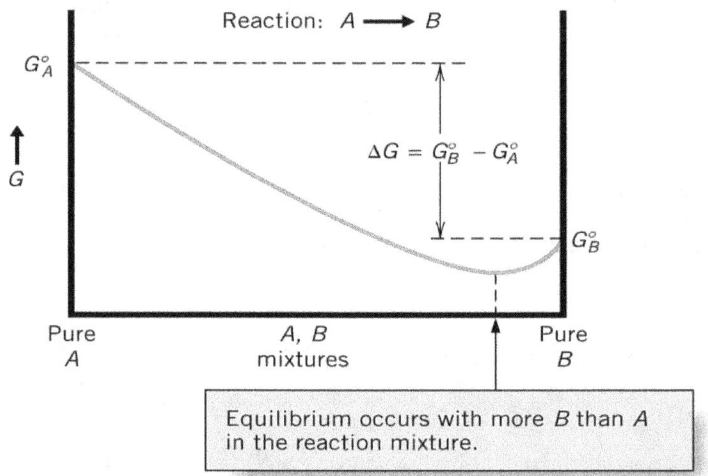

ΔG° and equilibrium constant K_c relationship

Equilibrium constant K_c is the ratio of the equilibrium concentrations of products over the equilibrium concentrations of reactants; each raised to the power of stoichiometric coefficients.

$$K_c = [\text{products}]^x / [\text{reactants}]^y$$

where [ ] represents concentration and exponents x and y are stoichiometric coefficients

For example, calculate the equilibrium constant K_c at 25 °C for the decarboxylation (~CO_2) of liquid pyruvic acid ($CH_3COCOOH$) to form gaseous acetaldehyde (CH_3COH) and carbon dioxide (CO_2).

$$CH_3COCOOH \;(l) \rightleftharpoons CH_3COH \;(g) + CO_2 \;(g)$$

Compound	ΔG°_f (kJ/mol)
CH_3COH	-33.30
$CH_3COCOOH$	-463.38
CO_2	-394.36

Equilibrium constant K_{eq} from ΔG

K_{eq} is a measure of the ratio of the concentrations of products to the concentrations of reactants.

ΔG° is standard change in free energy or the change in free energy *under standard conditions*.

$$\Delta G^\circ = -RT \ln K_{eq}$$

where ΔG° is the standard change in free energy, K_{eq} is the ratio of the concentrations of products to the concentrations of reactants, $R = 8.314$ J mol^{-1} K^{-1} and T is temperature in Kelvin

Sum ΔG°_f for each reaction to obtain total ΔG°.

$$\Delta G^\circ = \Sigma \Delta G^\circ_f$$

$$\Delta G^\circ = \Delta G^\circ_f \,(CH_3COH) + \Delta G^\circ_f \,(CO_2) - \Delta G^\circ_f \,(CH_3COCOOH)$$

$$\Delta G^\circ = (-133.30 \;(kJ/mol) + (-394.36 \;(kJ/mol) - (-463.38(kJ/mol))$$

$$\Delta G^\circ = -64.28 \text{ kJ/mol}$$

Use $\Delta G°$ value to calculate K_{eq}

$$K_{eq} = e^{-\Delta G°/RT}$$

where K is the equilibrium constant, R is the gas constant, and T is the temperature in Kelvin

Solve for exponent

$$\Delta G° / RT = (-64.28 \text{ kJ/mol}) / [(8.314 \text{ J/K}) \times (298 \text{ K}) \times (1{,}000 \text{ J/kJ})]$$

$$\Delta G° / RT = -25.94$$

Substitute exponent value

$$K_{eq} = e^{-(-25.945)}$$

$$K_{eq} = e^{25.945}$$

$$K_{eq} = 1.85 \times 10^{11}$$

Value for K_{eq} means that the ratio for concentration of products to reactants is 1.85×10^{11}

Therefore, product formation is favored.

Notes for active learning

Notes for active learning

Notes for active learning

PRACTICE QUESTIONS
&
DETAILED EXPLANATIONS

Practice Questions

Practice Set 1: Questions 1–20

1. What is the general equilibrium constant (K_{eq}) expression for the reversible reaction:

$$2\,A + 3\,B \leftrightarrow C$$

A. $K_{eq} = [C] / [A]^2 \cdot [B]^3$

B. $K_{eq} = [C] / [A] \cdot [B]$

C. $K_{eq} = [A] \cdot [B] / [C]$

D. $K_{eq} = [A]^2 \cdot [B]^3 / [C]$

E. none of the above

2. What are gases A and B likely to be, if in a mixture of these two gases, gas A has twice the average velocity of gas B?

A. Ar and Kr

B. N and Fe

C. Ne and Ar

D. Mg and K

E. B and Ne

3. For a hypothetical reaction, $A + B \rightarrow C$, predict which reaction occurs at the slowest rate from the following reaction conditions.

Reaction	Activation energy	Temperature
1	103 kJ/mol	15 °C
2	46 kJ/mol	22 °C
3	103 kJ/mol	24 °C
4	46 kJ/mol	30 °C

A. 1

B. 2

C. 3

D. 4

E. requires more information

4. From the data below, what is the order of the reaction with respect to reactant A?

Determining Rate Law from Experimental Data

$$A + B \rightarrow Products$$

Exp.	Initial [A]	Initial [B]	Initial Rate M/s
1	0.015	0.022	0.125
2	0.030	0.044	0.500
3	0.060	0.044	0.500
4	0.060	0.066	1.125
5	0.085	0.088	?

A. Zero

B. First

C. Second

D. Third

E. Fourth

5. For the combustion of ethanol (C_2H_6O) to form carbon dioxide and water, what is the rate at which carbon dioxide is produced if the ethanol is consumed at 4.0 M s^{-1}?

A. 1.5 M s^{-1}

B. 12.0 M s^{-1}

C. 8.0 M s^{-1}

D. 9.0 M s^{-1}

E. 10.0 M s^{-1}

6. Which is the correct equilibrium constant (K_{eq}) expression for the following reaction?

$$2\ Ag\ (s) + Cl_2\ (g) \leftrightarrow 2\ AgCl\ (s)$$

A. $K_{eq} = [AgCl] / [Ag]^2 \times [Cl_2]$

B. $K_{eq} = [2AgCl] / [2Ag] \times [Cl_2]$

C. $K_{eq} = [AgCl]^2 / [Ag]^2 \times [Cl_2]$

D. $K_{eq} = 1 / [Cl_2]$

E. $K_{eq} = [2AgCl]^2 / [2Ag]^2 \times [Cl_2]$

7. The equilibrium position for a system where $K_{eq} = 6.3 \times 10^{-14}$ can be described as being favored for [   ] and the concentration of products is relatively [   ].

A. the left; large

B. the left; small

C. the right; large

D. the right; small

E. neither direction; large

8. What can be deduced about the activation energy of a reaction that takes billions of years to go to completion and a reaction that takes only a fraction of a millisecond?

    A. The slow reaction has high activation energy, while the fast reaction has a low activation energy

    B. The slow reaction must have low activation energy, while the fast reaction must have a high activation energy

    C. The activation energy of both reactions is very low

    D. The activation energy of both reactions is very high

    E. The activation energy of both reactions is equal

9. Which influences the rate of a first-order chemical reaction?

    I. catalyst   II. temperature   III. concentration

    A. I only   **C.** I and III only

    B. I and II only   **D.** II and III only

                                  E. I, II and III

> Questions **10** through **14** are based on the following:

Energy profiles for four reactions (with the same scale).

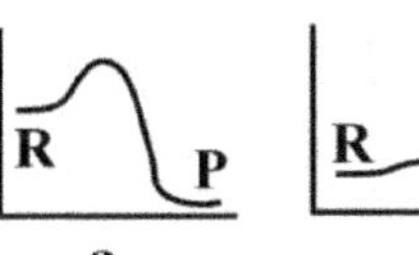

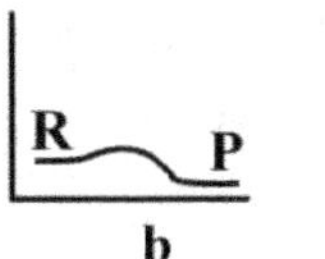

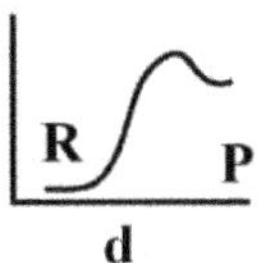

R = reactants P = products

10. Which reaction requires the most energy?

    A. a   **B.** b   **C.** c   **D.** d   **E.** None of the above

11. Which reaction has the highest activation energy?

    A. a   **B.** b   **C.** c   **D.** d   **E.** c and d

12. Which reaction has the lowest activation energy?

    A. a   **B.** b   **C.** c   **D.** d   **E.** a and b

13. Which reaction proceeds the slowest?

A. a **B.** b **C.** c **D.** d **E.** c and d

14. If the graphs are for the same reaction, which likely has a catalyst?

A. a
B. b

C. c
D. d
E. All have a catalyst

15. What is an explanation for observing that the reaction stops before all reactants are converted to products in the following reaction?

$$NH_3\,(aq) + HC_2H_3O_2\,(aq) \rightarrow NH_4^+\,(aq) + C_2H_3O_2^-\,(aq)$$

A. The catalyst is depleted
B. The reverse rate increases, while the forward rate decreases until they are equal
C. As [products] increases, the acetic acid begins to dissociate, stopping the reaction
D. As [reactants] decreases, NH_3 and $HC_2H_3O_2$ molecules stop colliding
E. As [products] increases, NH_3 and $HC_2H_3O_2$ molecules stop colliding

16. What is the term for the principle that the rate of reaction is regulated by the frequency, energy, and orientation of molecules striking each other?

A. orientation theory
B. frequency theory

C. energy theory
D. collision theory
E. rate theory

17. What is the effect on the energy of the activated complex and on the rate of the reaction when a catalyst is added to a chemical reaction?

A. The energy of the activated complex increases and the reaction rate decreases
B. The energy of the activated complex decreases and the reaction rate increases
C. The energy of the activated complex and the reaction rate increase
D. The energy of the activated complex and the reaction rate decrease
E. The energy of the activated complex remains the same, while the reaction rate decreases

18. Which change shifts the equilibrium to the right for the reversible reaction in an aqueous solution?

$$HNO_2\,(aq) \leftrightarrow H^+\,(aq) + NO_2^-\,(aq)$$

I. Add solid NaOH
II. Decrease $[NO_2^-]$
III. Decrease $[H^+]$
IV. Increase $[HNO_2]$

A. II and III only

B. II and IV only

C. III and IV only

D. II, III and IV only

E. I, II, III and IV

19. Which conditions would favor driving the reaction to completion?

$$2\,N_2\,(g) + 6\,H_2O\,(g) + heat \leftrightarrow 4\,NH_3\,(g) + 3\,O_2\,(g)$$

A. Increasing the reaction temperature

B. Continual addition of NH_3 gas to the reaction mixture

C. Decreasing the pressure on the reaction vessel

D. Continual removal of N_2 gas

E. Decreases reaction temperature

20. The system, $H_2\,(g) + X_2\,(g) \leftrightarrow 2\,HX\,(g)$ has a value of 24.4 for K_c. A catalyst was introduced into a reaction within a 4.0-liters reactor containing 0.20 moles of H_2, 0.20 moles of X_2 and 0.800 moles of HX. The reaction proceeds in which direction?

A. to the right, $Q > K_c$

B. to the left, $Q > K_c$

C. to the right, $Q < K_c$

D. to the left, $Q < K_c$

E. requires more information

Notes for active learning

Practice Set 2: Questions 21–40

21. Which is the K_c equilibrium expression for the following reaction?

$$4\ CuO\ (s) + CH_4\ (g) \leftrightarrow CO_2\ (g) + 4\ Cu\ (s) + 2\ H_2O\ (g)$$

A. $[Cu]^4 / [CuO]^4$

B. $[CH_4]^2 / [CO_2]\cdot[H_2O]$

C. $[CH_4] / [CO_2]\cdot[H_2O]^2$

D. $[CuO]^4 / [Cu]^4$

E. $[CO_2][H_2O]^2 / [CH_4]$

22. Which of the following concentrations of CH_2Cl_2 should be used in the rate law for Step 2 if CH_2Cl_2 is a product of the fast (first) step and a reactant of the slow (second) step?

A. $[CH_2Cl_2]$ at equilibrium

B. $[CH_2Cl_2]$ in Step 2 cannot be predicted because Step 1 is the fast step

C. Zero moles per liter

D. $[CH_2Cl_2]$ after Step 1 is completed

E. None of the above

23. What is the term for a substance that allows a reaction to proceed faster by lowering the energy of activation?

A. rate barrier

B. energy barrier

C. collision energy

D. activation energy

E. catalyst

24. Which statement is true for the grams of products present after a chemical reaction reaches equilibrium?

A. Must equal the grams of the initial reactants

B. May be less than, equal to, or greater than the grams of reactants present, depending upon the chemical reaction

C. Must be greater than the grams of the initial reactants

D. Must be less than the grams of the initial reactants

E. None of the above

25. What is the rate law when rates were measured at different concentrations for the dissociation of hydrogen gas: $H_2 (g) \rightarrow 2\,H (g)$?

$[H_2]$	Rate M/s s^{-1}
1.0	1.3×10^5
1.5	2.6×10^5
2.0	5.2×10^5

A. rate $= k^2[H] / [H_2]$

B. rate $= k[H]^2 / [H_2]$

C. rate $= k[H_2]^2$

D. rate $= k[H_2] / [H]^2$

E. requires more information

120. Which change to this reaction system causes the equilibrium to shift to the right?

$$N_2 (g) + 3\,H_2 (g) \leftrightarrow 2\,NH_3 (g) + heat$$

A. Heating the system

B. Removal of $H_2 (g)$

C. Addition of $NH_3 (g)$

D. Lowering the temperature

E. Addition of a catalyst

27. Which statement is NOT correct for $aA + bB \rightarrow dD + eE$ whereby rate $= k[A]^q \times [B]^r$?

A. The overall order of the reaction is $q + r$

B. The exponents q and r are equal to the coefficients a and b, respectively

C. The exponents q and r must be determined experimentally

D. The exponents q and r are often integers

E. The symbol k represents the rate constant

28. Which is the correct equilibrium constant (K_{eq}) expression for the following reaction?

$$CO (g) + 2\,H_2 (g) \leftrightarrow CH_3OH (l)$$

A. $K_{eq} = 1 / [CO] \cdot [H_2]^2$

B. $K_{eq} = [CO] \cdot [H_2]^2$

C. $K_{eq} = [CH_3OH] / [CO] \cdot [H_2]^2$

D. $K_{eq} = [CH_3OH] / [CO] \cdot [H_2]$

E. None of the above

Questions **29-32** are based on the following graph and net reaction:

The reaction proceeds in two consecutive steps.

$$XY + Z \leftrightarrow XYZ \leftrightarrow X + YZ$$

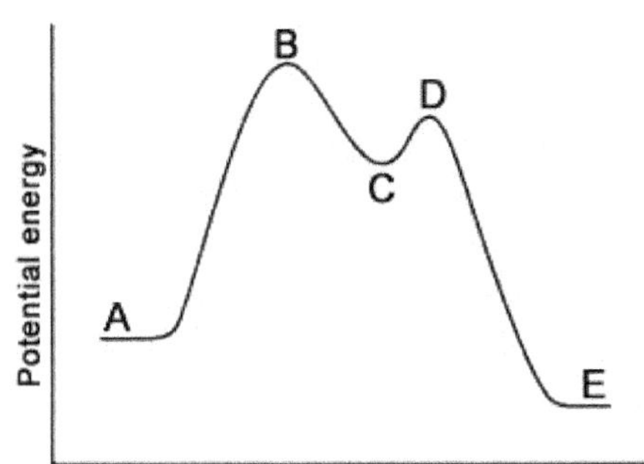

29. Where is the activated complex for this reaction?

A. A

B. A and C

C. C

D. E

E. B and D

30. The activation energy of the slow step for the forward reaction is given by:

A. A $\rightarrow$ B

B. A $\rightarrow$ C

C. B $\rightarrow$ C

D. C $\rightarrow$ E

E. A $\rightarrow$ E

31. The activation energy of the slow step for the reverse reaction is given by:

A. C $\rightarrow$ A

B. E $\rightarrow$ C

C. C $\rightarrow$ B

D. E $\rightarrow$ D

E. E $\rightarrow$ A

32. The change in energy (ΔE) of the overall reaction is given by the difference between:

A. A and B

B. A and E

C. A and C

D. B and D

E. B and C

33. Which of the following increases the collision energy of gaseous molecules?

> I. Increasing the temperature
>
> II. Adding a catalyst
>
> III. Increasing the concentration

A. I only

B. II only

C. III only

D. I and II only

E. I and III only

34. Which of the following conditions favors the formation of NO (g) in a closed container?

$$N_2\ (g) + O_2\ (g) \leftrightarrow 2\ NO\ (g)$$

(Use $\Delta H = +181$ kJ/mol)

A. Increasing the temperature

B. Decreasing the temperature

C. Increasing the pressure

D. Decreasing the pressure

E. Decreasing the temperature and decreasing the pressure

35. A chemical system is considered to have reached dynamic equilibrium when the:

A. activation energy of the forward reaction equals the activation energy of the reverse reaction

B. rate of production of each of the products equals the rate of their consumption by the reverse reaction

C. frequency of collisions between the reactant molecules equals the frequency of collisions between the product molecules

D. sum of the concentrations of each of the reactant species equals the sum of the concentrations of each of the product species

E. none of the above

36. Chemical equilibrium is reached in a system when:

A. complete conversion of reactants to products has occurred

B. product molecules begin reacting with each other

C. reactant concentrations steadily decrease

D. reactant concentrations steadily increase

E. product and reactant concentrations remain constant

37. At a given temperature, $K = 46.0$ for the reaction:

$$4\ HCl\ (g) + O_2\ (g) \leftrightarrow 2\ H_2O\ (g) + 2\ Cl_2\ (g)$$

At equilibrium, $[HCl] = 0.150$, $[O_2] = 0.395$ and $[H_2O] = 0.625$. What is the concentration of Cl_2 at equilibrium?

A. 0.153 M

B. 0.444 M

C. 1.14 M

D. 0.00547 M

E. 2.64 M

38. If at equilibrium, reactant concentrations are slightly smaller than product concentrations, the equilibrium constant would be:

A. slightly greater than 1

B. slightly lower than 1

C. much lower than 1

D. much greater than 1

E. equal to zero

39. Hydrogen gas reacts with iron (III) oxide to form iron metal (which produces steel), as shown in the reaction below. Which statement is NOT correct concerning the equilibrium system?

$$Fe_2O_3\ (s) + 3\ H_2\ (g) + heat \leftrightarrow 2\ Fe\ (s) + 3\ H_2O\ (g)$$

A. Continually removing water from the reaction chamber increases the yield of iron

B. Decreasing the volume of hydrogen gas reduces the yield of iron

C. Lowering the reaction temperature increases the concentration of hydrogen gas

D. Increasing the pressure on the reaction chamber increases the formation of products

E. Decreasing the reaction temperature reduces the formation of iron

40. Which changes shift the equilibrium to the right for the reversible reaction:

$$CO\ (g) + H_2O\ (g) \leftrightarrow CO_2\ (g) + H_2\ (g) + heat$$

A. increasing volume

B. increasing temperature

C. increasing $[CO_2]$

D. adding a catalyst

E. increasing $[CO]$

Notes for active learning

Notes for active learning

Practice Set 3: Questions 41–60

41. Which is the correct equilibrium constant (K_{eq}) expression for the following reaction?

$$4\ NH_3\ (g) + 5\ O_2\ (g) \leftrightarrow 4\ NO\ (g) + 6\ H_2O\ (g)$$

A. $K_{eq} = [NO]^4 \times [H_2O]^6 / [NH_3]^4 \times [O_2]^5$
B. $K_{eq} = [NH_3]^4 \times [O_2]^5 / [NO]^4 \times [H_2O]^6$
C. $K_{eq} = [NO] \times [H_2O] / [NH_3] \times [O_2]$
D. $K_{eq} = [NH_3] \times [O_2] / [NO] \times [H_2O]$
E. $K_{eq} = [NH_3]^2 \times [O_2]^5 / [NO]^2 \times [H_2O]^3$

42. Carbonic acid equilibrium in blood:

$$CO_2\ (g) + H_2O\ (l) \leftrightarrow H_2CO_3\ (aq) \leftrightarrow H^+\ (aq) + HCO_3^-\ (aq)$$

If a person hyperventilates, rapid breathing expels carbon dioxide gas. Which of the following decreases when a person hyperventilates?

I. $[HCO_3^-]$ II. $[H^+]$ III. $[H_2CO_3]$

A. I only
B. II only
C. III only
D. I and II only
E. I, II and III

43. What is the overall order of the reaction if the units of the rate constant for a particular reaction are min^{-1}?

A. Zero
B. First
C. Second
D. Third
E. Fourth

44. Heat is often added to chemical reactions performed in the laboratory to:

A. compensate for the natural tendency of energy to disperse
B. increase the rate at which reactants collide
C. allow a greater number of reactants to overcome the barrier of the activation energy
D. increase the energy of the reactant molecules
E. all of the above

45. Predict which reaction occurs at a faster rate for a hypothetical reaction X + Y → W + Z.

Reaction	Activation energy	Temperature
1	low	low
2	low	high
3	high	high
4	high	low

A. 1   **B.** 2   **C.** 3   **D.** 4   **E.** 1 and 4

46. For the reaction, $2\,XO + O_2 \rightarrow 2\,XO_2$, data obtained from measurement of the initial rate of reaction at varying concentrations are:

Experiment	[XO]	[O$_2$]	Rate (mmol L^{-1} s^{-1})
1	0.010	0.010	2.5
2	0.010	0.020	5.0
3	0.030	0.020	45.0

What is the expression for the rate law?

A. rate = $k[XO]\cdot[O_2]$

B. rate = $k[XO]^2\cdot[O_2]^2$

C. rate = $k[XO]^2\cdot[O_2]$

D. rate = $k[XO]\cdot[O_2]^2$

E. rate = $k[XO]^2 / [O_2]^2$

47. What is the equilibrium (K_{eq}) expression for the following reaction?

$$CaO\ (s) + CO_2\ (g) \leftrightarrow CaCO_3\ (s)$$

A. $K_{eq} = [CaCO_3] / [CaO]$

B. $K_{eq} = 1 / [CO_2]$

C. $K_{eq} = [CaCO_3] / [CaO]\cdot[CO_2]$

D. $K_{eq} = [CO_2]$

E. $K_{eq} = [CaO]\cdot[CO_2] / [CaCO_3]$

48. If a reaction does not occur extensively and gives a low concentration of products at equilibrium, which of the following is true?

A. The rate of the forward reaction is greater than the reverse reaction

B. The rate of the reverse reaction is greater than the forward reaction

C. The equilibrium constant is greater than one; that is, K_{eq} is larger than 1

D. The equilibrium constant is less than one; that is, K_{eq} is smaller than 1

E. The equilibrium constant equals 1

49. Which of the following changes most likely decreases the rate of a reaction?

   A. Increasing the reaction temperature

   B. Increasing the concentration of a reactant

   C. Increasing the activation energy for the reaction

   D. Decreasing the activation energy for the reaction

   E. Increasing the reaction pressure

50. Which factors would increase the rate of a reversible chemical reaction?

   I. Increasing the temperature of the reaction

   II. Removing products as they form

   III. Adding a catalyst to the reaction vessel

   A. I only   **C.** I and II only

   B. II only   **D.** I and III only

                              E. I, II and III

51. What is the ionization equilibrium constant (K_i) expression for the following weak acid?

$$H_2S\,(aq) \leftrightarrow H^+\,(aq) + HS^-\,(aq)$$

   A. $K_i = [H^+]^2 \cdot [S^{2-}] / [H_2S]$ **C.** $K_i = [H^+] \cdot [HS^-] / [H_2S]$

   B. $K_i = [H_2S] / [H^+] \cdot [HS^-]$ **D.** $K_i = [H^+]^2 \cdot [HS^-] / [H_2S]$

                              E. $K_i = [H_2S] / [H^+]^2 \cdot [HS^-]$

52. Increasing the temperature of a chemical reaction:

   A. increases the reaction rate by lowering the activation energy

   B. increases the reaction rate by increasing reactant collisions per unit time

   C. increases the activation energy, thus increasing the reaction rate

   D. raises the activation energy, thus decreasing the reaction rate

   E. causes fewer reactant collisions to take place

53. All of the following factors determine reaction rates, EXCEPT:

   A. orientation of collisions between molecules

   B. spontaneity of the reaction

   C. force of collisions between molecules

   D. number of collisions between molecules

   E. the activation energy of the reaction

54. Coal burning plants release sulfur dioxide into the atmosphere, while nitrogen monoxide is released via industrial processes and combustion engines. Sulfur dioxide can be produced in the atmosphere by the following equilibrium reaction:

$$SO_3 \, (g) + NO \, (g) + heat \leftrightarrow SO_2 \, (g) + NO_2 \, (g)$$

Which of the following does NOT shift the equilibrium to the right?

A. $[NO_2]$ decrease

B. $[NO]$ increase

C. Decrease the reaction chamber volume

D. Temperature increase

E. All of the above shift the equilibrium to the right

55. Which of the following statements can be assumed to be true about how reactions occur?

A. Reactant particles must collide with each other

B. Energy must be released as the reaction proceeds

C. Catalysts must be present in the reaction

D. Energy must be absorbed as the reaction proceeds

E. The energy of activation must have a negative value

56. At equilibrium, increasing the temperature of an exothermic reaction likely:

A. increases the heat of reaction

B. decreases the heat of reaction

C. increases the forward reaction

D. decreases the forward reaction

E. increases the heat of reaction and the forward reaction

57. Which shifts the equilibrium to the left for the reversible reaction in an aqueous solution?

$$HC_2H_3O_2 \, (aq) \leftrightarrow H^+ \, (aq) + C_2H_3O_2^- \, (aq)$$

I. increase pH

II. increase $[HC_2H_3O_2]$

III. add solid $KC_2H_3O_2$

A. I only

B. I and III only

C. II only

D. II and III only

E. III only

58. Which of the following statements is true concerning the equilibrium system, whereby S combines with H_2 to form hydrogen sulfide, toxic gas from the decay of organic material? (Use the equilibrium constant, $K_{eq} = 2.8 \times 10^{-21}$)

$$S\,(g) + H_2\,(g) \leftrightarrow H_2S\,(g)$$

- **A.** Almost all the starting molecules are converted to product
- **B.** Decreasing $[H_2S]$ shifts the equilibrium to the left
- **C.** Decreasing $[H_2]$ shifts the equilibrium to the right
- **D.** Increasing the volume of the sealed reaction container shifts the equilibrium to the right
- **E.** Very little hydrogen sulfide gas is present in the equilibrium

59. Which of the conditions characterizes a system in a state of chemical equilibrium?

- **A.** Product concentrations are greater than reactant concentrations
- **B.** Reactant molecules no longer react with each other
- **C.** Concentrations of reactants and products are equal
- **D.** The rate of the forward reaction has dropped to zero
- **E.** Reactants are being consumed at the same rate they are being produced

60. Which statement is NOT true regarding an equilibrium constant for a particular reaction?

- **A.** It does not change as the product is removed
- **B.** It does not change as an additional quantity of a reactant is added
- **C.** It changes when a catalyst is added
- **D.** It changes as the temperature increases
- **E.** All are true statements

Notes for active learning

Practice Set 4: Questions 61–80

Questions **61** through **63** refer to the rate data
for the conversion of reactants W, X, and Y to product Z.

Trial Number	Concentration (moles/L)			Rate of Formation of Z (moles/l·s)
	W	X	Y	
1	0.01	0.05	0.04	0.04
2	0.015	0.07	0.06	0.08
3	0.01	0.15	0.04	0.36
4	0.03	0.07	0.06	0.08
5	0.01	0.05	0.16	0.08

61. From the above data, what is the overall order of the reaction?

A. $3\frac{1}{2}$ **B.** 4   **C.** 3   **D.** 2   **E.** $2\frac{1}{2}$

62. From the above data, the order with respect to W suggests that the rate of formation of Z is:

A. dependent on [W]

B. independent of [W]

C. semi-dependent on [W]

D. unable to be determined

E. inversely proportional to [W]

63. From the above data, the magnitude of *k* for trial 1 is:

A. 20   **B.** 40   **C.** 60   **D.** 80   **E.** 90

64. Which of the following changes shifts the equilibrium to the left for the given reversible reaction?

$$SO_3 (g) + NO (g) + heat \leftrightarrow SO_2 (g) + NO_2 (g)$$

A. Decrease temperature

B. Decrease volume

C. Increase [NO]

D. Decrease [SO_2]

E. Add a catalyst

---

65. What is the ionization equilibrium constant (K_i) expression for the following weak acid?

$$H_3PO_4 \,(aq) \leftrightarrow H^+ (aq) + H_2PO_4^- \,(aq)$$

A. $K_i = [H_3PO_4] / [H^+] \cdot [H_2PO_4^-]$

B. $K_i = [H^+]^3 \cdot [PO_4^{3-}] / [H_3PO_4]$

C. $K_i = [H^+]^3 \cdot [H_2PO_4^-] / [H_3PO_4]$

D. $K_i = [H^+] \cdot [H_2PO_4^-] / [H_3PO_4]$

E. $K_i = [H_3PO_4] / [H^+]^3 \cdot [PO_4^{3-}]$

66. What effect does a catalyst have on equilibrium?

A. It increases the rate of the forward reaction

B. It shifts the reaction to the right

C. It increases the rate at which equilibrium is reached without changing ΔG

D. It increases the rate at which equilibrium is reached and lowers ΔG

E. It slows the reverse reaction

67. What is the ionization equilibrium constant (K_i) expression for the following weak acid?

$$H_2SO_3 \,(aq) \leftrightarrow H^+ (aq) + HSO_3^- \,(aq)$$

A. $K_i = [H_2SO_3] / [H^+] \cdot [HSO_3^-]$

B. $K_i = [H^+]^2 \cdot [SO_3^{2-}] / [H_2SO_3]$

C. $K_i = [H^+]^2 \cdot [HSO_3^-] / [H_2SO_3]$

D. $K_i = [H^+] \cdot [HSO_3^-] / [H_2SO_3]$

E. $K_i = [H_2SO_3] / [H^+]^2 \cdot [SO_3^{2-}]$

68. Which of the following is true if a reaction occurs extensively and yields a high concentration of products at equilibrium?

A. The rate of the reverse reaction is greater than the forward reaction

B. The rate of the forward reaction is greater than the reverse reaction

C. The equilibrium constant is less than one; K_{eq} is much smaller than 1

D. The equilibrium constant is greater than one; K_{eq} is much larger than 1

E. The equilibrium constant equals 1

69. The minimum combined kinetic energy reactants must possess for collisions to result in a reaction is:

A. orientation energy

B. activation energy

C. collision energy

D. dissociation energy

E. bond energy

70. Which factors decrease the rate of a reaction?

> I. Lowering the temperature
> II. Increasing the concentration of reactants
> III. Adding a catalyst to the reaction vessel

A. I only

B. II only

C. III only

D. I and II only

E. I, II and III

71. For a collision between molecules to result in a reaction, the molecules must possess a favorable orientation relative to each other and:

A. be in the gaseous state

B. have a certain minimum energy

C. adhere for at least 2 nanoseconds

D. exchange electrons

E. be in the liquid state

72. Most reactions are carried out in a liquid solution or the gaseous phase, because in such situations:

A. kinetic energies of reactants are lower

B. reactant collisions occur more frequently

C. activation energies are higher

D. reactant activation energies are lower

E. reactant collisions occur less frequently

73. Find the reaction rate for $A + B \rightarrow C$:

Trial	$[A]_{t=0}$	$[B]_{t=0}$	Initial rate (M/s)
1	0.05 M	1.0 M	1.0×10^{-3}
2	0.05 M	4.0 M	16.0×10^{-3}
3	0.15 M	1.0 M	3.0×10^{-3}

A. rate $= k[A]^2 \cdot [B]^2$

B. rate $= k[A] \cdot [B]^2$

C. rate $= k[A]^2 \cdot [B]$

D. rate $= k[A] \cdot [B]$

E. rate $= k[A]^2 \cdot [B]^3$

74. Why does a glowing splint of wood burn only slowly in air, but rapidly in a burst of flames when placed in pure oxygen?

- **A.** A glowing wood splint is extinguished within pure oxygen because oxygen inhibits the smoke
- **B.** Pure oxygen can absorb carbon dioxide at a faster rate
- **C.** Oxygen is a flammable gas
- **D.** There is an increased number of collisions between the wood and oxygen molecules
- **E.** There is a decreased number of collisions between the wood and oxygen

75. Which change shifts the equilibrium to the right for the following system at equilibrium?

$$N_2(g) + 3 H_2(g) \leftrightarrow 2 NH_3(g) + 92.94 \text{ kJ}$$

I. Removing NH_3

II. Adding NH_3

III. Removing N_2

IV. Adding N_2

- **A.** I and III
- **B.** II and III
- **C.** II and IV
- **D.** I and IV
- **E.** None of the above

76. Which of the changes does not affect the equilibrium for the reversible reaction in an aqueous solution?

$$HC_2H_3O_2(aq) \leftrightarrow H^+(aq) + C_2H_3O_2^-(aq)$$

- **A.** Adding solid $NaC_2H_3O_2$
- **B.** Adding solid $NaNO_3$
- **C.** Increasing $[HC_2H_3O_2]$
- **D.** Increasing $[H^+]$
- **E.** Adding solid NaOH

77. Consider the following reaction: $H_2(g) + I_2(g) \rightarrow 2 HI(g)$

At 160 K, this reaction has an equilibrium constant of 35. If at 160 K, the concentration of hydrogen gas is 0.4 M, iodine gas is 0.6 M, and hydrogen iodide gas is 3 M:

- **A.** system is at equilibrium
- **B.** [iodine] decreases
- **C.** [hydrogen iodide] increases
- **D.** [hydrogen iodide] decreases
- **E.** [hydrogen] decreases

78. If the concentration of reactants decreases, which of the following is true?

    I. The amount of products increases

    II. The heat of reaction decreases

    III. The rate of reaction decreases

A. I only

B. II only

C. III only

D. II and III only

E. I, II and III

79. What does a chemical equilibrium expression of a reaction depend on?

    I. mechanism

    II. stoichiometry

    III. rate

A. I only

B. II only

C. III only

D. I and II only

E. I, II and III

80. For the following reaction where $\Delta H < 0$, which factor decreases the magnitude of the equilibrium constant K?

$$CO\ (g) + 2\ H_2O\ (g) \leftrightarrow CH_3OH\ (g)$$

A. Decreasing the temperature of this system

B. Decreasing volume

C. Decreasing the pressure of this system

D. All of the above

E. None of the above

Notes for active learning

Practice Set 5: Questions 81–100

81. Which changes shift the equilibrium to the product side for the following reaction at equilibrium?

$$SO_2Cl_2\ (g) \leftrightarrow SO_2\ (g) + Cl_2\ (g)$$

I. Addition of SO_2Cl_2
II. Addition of SO_2
III. Removal of SO_2Cl_2
IV. Removal of Cl_2

A. I, II and III
B. II and III
C. I and II
D. III and IV
E. I and IV

82. Calculate a value for K_c for the following reaction:

$$NOCl\ (g) + \tfrac{1}{2}\,O_2\ (g) \leftrightarrow NO_2\ (g) + \tfrac{1}{2}\,Cl_2\ (g)$$

Use the data:

$$2\,NO\ (g) + Cl_2\ (g) \leftrightarrow 2\,NOCl\ (g) \qquad K_c = 3.20 \times 10^{-3}$$

$$2\,NO_2\ (g) \leftrightarrow 2\,NO\ (g) + O_2\ (g) \qquad K_c = 15.5$$

A. 4.49
B. 0.343
C. 4.32×10^{-4}
D. 1.33×10^{-5}
E. 18.4

83. Assuming vessels a, b and c are drawn to relative proportions, which of the following reactions proceeds the fastest? (Assume equal temperatures)

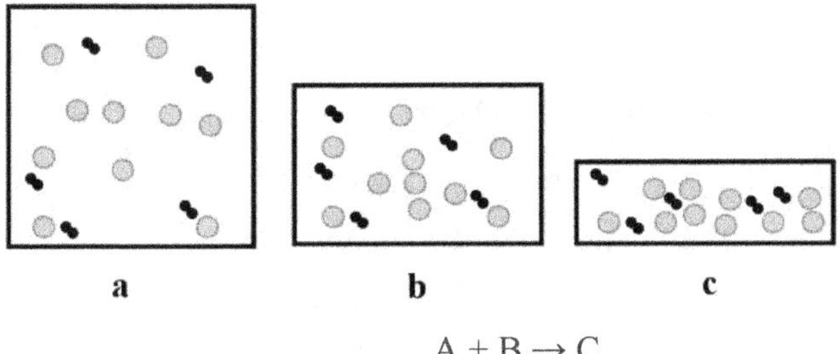

A. a
B. b
C. c
D. All proceed at same rate
E. Not enough information given

84. The reaction rate is:

A. the ratio of the masses of products and reactants

B. the ratio of the molecular masses of the elements in a given compound

C. the speed at which reactants are consumed or product is formed

D. the balanced chemical formula relating the number of product to reactant molecules

E. none of the above

85. What is the equilibrium constant (K_{eq}) expression for the reversible reaction below?

$$2\,A \leftrightarrow B + 3\,C$$

A. $K_{eq} = [B] \times [C]^3 / [A]^2$

B. $K_{eq} = [B] \times [C] / [A]$

C. $K_{eq} = [A]^2 / [B] \times [C]^3$

D. $K_{eq} = [A] / [B] \times [C]$

E. None of the above

86. With respect to A, what is the order of the reaction if the rate law $= k[A]^3[B]^6$?

A. 2   B. 3   C. 4   D. 5   E. 6

87. Nitrogen monoxide reacts with bromine at elevated temperatures according to the equation:

$$2\,NO\,(g) + Br_2\,(g) \rightarrow 2\,NOBr\,(g)$$

What is the rate of consumption of $Br_2\,(g)$ if, in a certain reaction mixture, the rate of formation of $NOBr\,(g)$ was 4.50×10^{-4} mol L^{-1} s^{-1}?

A. 3.12×10^{-4} mol L^{-1} s^{-1}

B. 8.00×10^{-4} mol L^{-1} s^{-1}

C. 2.25×10^{-4} mol L^{-1} s^{-1}

D. 4.50×10^{-4} mol L^{-1} s^{-1}

E. 4.00×10^{-3} mol L^{-1} s^{-1}

88. Which of the following statements about "activation energy" is correct?

A. Activation energy is the energy given off when reactants collide

B. Activation energy is high for reactions that occur rapidly

C. Activation energy is low for reactions that occur rapidly

D. Activation energy is the maximum energy a reacting molecule may possess

E. Activation energy is low for reactions that occur slowly

89. Which is the correct equilibrium constant (K_{eq}) expression for the following reaction?

$$A + 2\,B \leftrightarrow 2\,C + D$$

A. $[C]^2 \cdot [D] / [A] \cdot [B]^2$

B. $[A]^2 \cdot [B]^2 / 2[C]^2 \cdot [D]$

C. $[A] \cdot 2[B] / 2[C] \cdot [D]$

D. $2[C] \cdot [D] / [A] \cdot 2[B]$

E. $[A] \cdot [B]^2 / [C]^2 \cdot [D]$

90. What is the effect on the equilibrium after adding H_2O to the equilibrium mixture if CO_2 and H_2 react until equilibrium is established?

$$CO_2\ (g) + H_2\ (g) \leftrightarrow H_2O\ (g) + CO\ (g)$$

A. $[H_2]$ decreases and $[H_2O]$ increases

B. $[CO]$ and $[CO_2]$ increase

C. $[H_2]$ decreases and $[CO_2]$ increases

D. Equilibrium shifts to the left

E. Equilibrium shifts to the right

91. For a reaction that has an equilibrium constant of 4.3×10^{-17} at 25 °C, the relative position of equilibrium is described as:

A. equal amounts of reactants and products

B. significant amounts of both reactants and products

C. mostly products

D. mostly reactants

E. amount of product is slightly greater than reactants

92. What is the term for the energy necessary for reactants to achieve the transition state and form products?

A. heat of reaction

B. energy barrier

C. rate barrier

D. collision energy

E. activation energy

93. What is the equilibrium constant K_c value if, at equilibrium, the concentrations of $[NH_3]$ − 0.40 M, $[H_2]$ − 0.12 M and $[N_2]$ − 0.040 M?

$$2\,NH_3\ (g) \leftrightarrow N_2\ (g) + 3\,H_2\ (g)$$

A. 6.3×10^{12}

B. 7.1×10^{-7}

C. 4.3×10^{-4}

D. 3.9×10^{-3}

E. 8.5×10^{-9}

94. Which of the following increases the collision frequency of molecules?

    I. Increasing the concentration

    II. Adding a catalyst

    III. Decreasing the temperature

A. I only

B. II only

C. III only

D. I and II only

E. I and III only

95. For the following reaction with $\Delta H < 0$, which factor increases the equilibrium yield for methanol?

$$CO\ (g) + 2\ H_2O\ (g) \leftrightarrow CH_3OH\ (g)$$

    I. Decreasing the volume

    II. Decreasing the pressure

    III. Lowering the temperature of the system

A. I only

B. II only

C. I and II only

D. I and III only

E. I, II and III

96. If there is too much chlorine in the water, swimmers complain that their eyes burn. Consider the equilibrium found in swimming pools. Predict which increases the chlorine concentration.

$$Cl_2\ (g) + H_2O\ (l) \leftrightarrow HClO\ (aq) \leftrightarrow H^+\ (aq) + ClO^-\ (aq)$$

A. Decreasing the pH

B. Adding hydrochloric acid, HCl (aq)

C. Adding hypochlorous acid, HClO (aq)

D. Adding sodium hypochlorite, NaClO

E. All of the above

97. What is the ratio of the diffusion rates of H_2 gas to O_2 gas?

A. 4:1

B. 2:1

C. 1:3

D. 1:4

E. 1:2

98. Which of the following increases the amount of product formed from a reaction?

    I. Using a UV light catalyst

    II. Adding an acid catalyst

    III. Adding a metal catalyst

A. I only

B. II only

C. III only

D. I and II only

E. None of the above

99. A mixture of 1.40 moles of A and 2.30 moles of B reacted. At equilibrium, 0.90 moles of A are present. How many moles of C is present at equilibrium?

$$3 \text{ A } (g) + 2 \text{ B } (g) \rightarrow 4 \text{ C } (g)$$

A. 2.7 moles

B. 1.3 moles

C. 0.09 moles

D. 1.8 moles

E. 0.67 moles

100. What is the equilibrium constant K_c for the following reaction?

$$PCl_5 (g) + 2 \text{ NO } (g) \leftrightarrow PCl_3 (g) + 2 \text{ NOCl } (g)$$

$$K_1 = PCl_3 (g) + Cl_2 (g) \leftrightarrow PCl_5 (g)$$

$$K_2 = 2 \text{ NO } (g) + Cl_2 (g) \leftrightarrow 2 \text{ NOCl } (g)$$

A. K_1 / K_2

B. $(K_1 K_2)^{-1}$

C. $K_1 \times K_2$

D. K_2 / K_1

E. $K_2 - K_1$

Notes for active learning

Practice Set 6: Questions 101–120

101. Write the mass action (K_c) expression for the following reaction?

$$4 \text{ Cr } (s) + 3 \text{ CCl}_4 (g) \leftrightarrow 4 \text{ CrCl}_3 (g) + 4 \text{ C } (s)$$

A. $K_c = [\text{C}] \cdot [\text{CrCl}_3] / [\text{Cr}] \cdot [\text{CCl}_4]$

B. $K_c = [\text{C}]^4 \cdot [\text{CrCl}_3]^4 / [\text{Cr}]^4 \cdot [\text{CCl}_4]^3$

C. $K_c = [\text{CrCl}_3]^4 / [\text{CCl}_4]^3$

D. $K_c = [\text{CrCl}_3] / [\text{CCl}_4]$

E. $K_c = [\text{CrCl}_3] + [\text{CCl}_4]$

102. Which of the changes shifts the equilibrium to the left for the reversible reaction in an aqueous solution?

$$\text{HNO}_2 (aq) \leftrightarrow \text{H}^+ (aq) + \text{NO}_2^- (aq)$$

A. Adding solid KNO_2

B. Adding solid KCl

C. Increasing $[\text{HNO}_2]$

D. Increasing pH

E. Adding solid KOH

103. Given the equation $x\text{A} + y\text{B} \rightarrow z$, the rate expression reaction in terms of the rate of change of the concentration with respect to time is:

A. $k[\text{A}]^x[\text{B}]^{yz}$

B. $k[\text{A}]^y[\text{B}]^x$

C. $k[\text{A}]^x[\text{B}]^y$

D. $k[\text{A}]^x[\text{B}]^y / [z]$

E. cannot be determined

104. What is the rate law for the reaction $3 \text{ D} + \text{E} \rightarrow \text{F} + 2 \text{ G}$, given the following experimental data?

Experiment	[D]	[E]	Rate (mol L^{-1} s^{-1})
1	0.100	0.250	0.000250
2	0.200	0.250	0.000500
3	0.100	0.500	0.00100

A. rate $= k[\text{D}]^2 \cdot [\text{E}]^2$

B. rate $= k[\text{D}]^2 \cdot [\text{E}]$

C. rate $= k[\text{D}]^3 \cdot [\text{E}]$

D. rate $= k[\text{D}] \cdot [\text{E}]^2$

E. rate $= k[\text{D}] \cdot [\text{E}]$

105. A 10-mm cube of copper metal is placed in 400 mL of 10 M nitric acid at 27 °C and the reaction below occurs:

$$Cu(s) + 4\ H^+(aq) + 2\ NO_3^-(aq) \rightarrow Cu^{2+}(aq) + 2\ NO_2(g) + 2\ H_2O(l)$$

If nitrogen dioxide is produced at the rate of $2.8 \times 10{-}4$ M/min, what is the rate at which hydrogen ions are consumed?

A. 3.2×10^{-3} M/min

B. 1.8×10^{-2} M/min

C. 1.1×10^{-4} M/min

D. 5.6×10^{-4} M/min

E. 6.7×10^{-5} M/min

106. What is the equilibrium constant (K_{eq}) expression for the following reaction?

$$2\ CO(g) + O_2(g) \leftrightarrow 2\ CO_2(g)$$

A. $K_{eq} = 2[CO_2] / 2[CO] \times [O_2]$

B. $K_{eq} = [CO] \times [O_2] / [CO_2]$

C. $K_{eq} = [CO_2]^2 / [CO]^2 \times [O_2]$

D. $K_{eq} = [CO_2] / [CO] + [O_2]$

E. $K_{eq} = [CO]^2 \times [O_2] / [CO_2]^2$

107. In reaction $A + B \rightarrow AB$, which does NOT increase the reaction rate?

I. increasing the temperature

II. decreasing the temperature

III. adding a catalyst

A. I only

B. II only

C. III only

D. I and III only

E. II and III only

108. If $K_{eq} = 6.1 \times 10^{-11}$, which statement is true?

A. Slightly more products are present

B. The number of reactants equals products

C. Mostly products are present

D. Mostly reactants are present

E. Cannot be determined

109. What is the equilibrium constant for step 1 in a reaction by convention?

A. $k_1 k_{+2}$

B. $k_1 + k_{+2}$

C. k_{-1}

D. k_1

E. k_1 / k_{-1}

110. A catalyst changes which of the following?

A. ΔS

B. ΔG

C. ΔH

D. $E_{activation}$

E. $\Sigma(\Delta H_{products})$

111. Which changes shift(s) the equilibrium to the right for the reversible reaction in an aqueous solution?

$$HC_2H_3O_2\ (aq) \leftrightarrow H^+\ (aq) + C_2H_3O_2^-\ (aq)$$

I. Decreasing $[C_2H_3O_2^-]$

II. Decreasing $[H^+]$

III. Increasing $[HC_2H_3O_2]$

IV. Decreasing $[HC_2H_3O_2]$

A. I and II only

B. I and III only

C. II and III only

D. I, II and III only

E. I, II and IV only

112. For the reaction $A + B \rightarrow C$, which proceeds the slowest? (Assume equal temperatures)

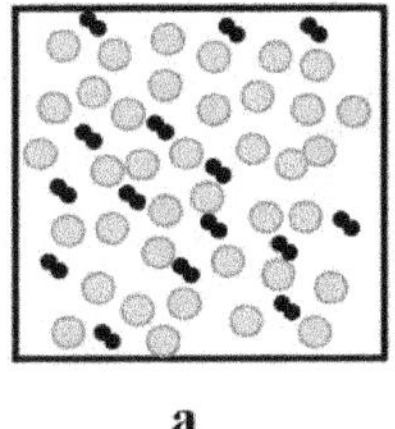
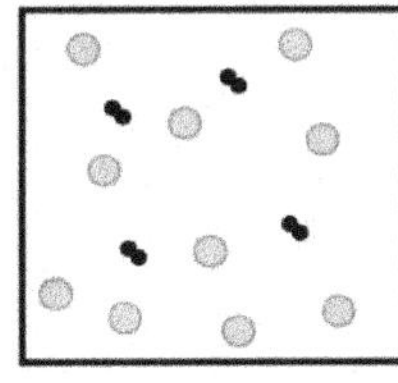

a   b   c

A. a

B. b

C. c

D. All proceed at the same rate

E. Not enough information

113. Why might increasing the concentration of a set of reactants increase the reaction rate?

A. The rate of reaction depends only on the mass of the atoms and increases as the mass of the reactants increase

B. There is an increased probability that any two reactant molecules collide and react

C. There is an increased ratio of reactants to products

D. The concentration of reactants is unrelated to the rate of reaction

E. None of the above

114. For a chemical reaction to occur, all of the following must happen, EXCEPT:

  A. reactant particles must collide with the correct orientation

  B. a large enough number of collisions must occur

  C. chemical bonds must break or form

  D. reactant particles must collide with enough energy for change to occur

  E. none of the above

115. If a catalyst is added to the reaction, which direction does the equilibrium shift toward?

$$CO + H_2O + \textit{heat} \leftrightarrow CO_2 + H_2$$

  A. To the left

  B. To the right

  C. No effect

  D. Not enough information

  E. Initially to the right, but settles at a ½ current equilibrium

116. For a chemical reaction at equilibrium, which decreases the concentration of products?

  A. Decreasing the pressure

  B. Increasing the temperature and decreasing the temperature

  C. Increasing the temperature

  D. Decreasing the temperature

  E. Decreasing the concentration of a gaseous or aqueous reactant

117. Which factor does NOT describe activated complexes?

  A. May be chemically isolated

  B. Decompose rapidly

  C. Have specific geometry

  D. Are extremely reactive

  E. At the high point of the reaction profile

118. When a reaction system is at equilibrium:

  A. rate in the forward and reverse directions is equal

  B. rate in the forward direction is at a maximum

  C. amounts of reactants and products are equal

  D. rate in the reverse direction is at a minimum

  E. reaction is complete and static

119. A reaction vessel contains NH_3, N_2, and H_2 at equilibrium with $[NH_3] = 0.1$ M, $[N_2] = 0.2$ M, and $[H_2] = 0.3$ M. For decomposition, what is K for $NH_3 \rightarrow N_2$ and H_2?

A. $K = (0.2) \cdot (0.3)^3 / (0.1)^2$

B. $K = (0.1) / (0.2)^2 \cdot (1.5)^3$

C. $K = (0.1)^2 / (0.2) \cdot (0.3)^3$

D. $K = (0.2) \cdot (0.3)^2 / (0.1)$

E. $K = (0.2) \cdot (0.3)^3 / (0.1)$

120. The data provides the rate of a reaction as affected by the concentration of the reactants. What is the order of the reaction rate?

Experiment	[X]	[Y]	[Z]	Rate (mol L^{-1} hr^{-1})
1	0.200 M	0.100 M	0.600 M	5.0
2	0.200 M	0.400 M	0.400 M	80.0
3	0.600 M	0.100 M	0.200 M	15.0
4	0.200 M	0.100 M	0.200 M	5.0
5	0.200 M	0.200 M	0.400 M	20.0

A. Zero order with respect to X

B. Order for X is minus one (rate proportional to 1 / [X])

C. First order with respect to X

D. Second order with respect to X

E. Order for X cannot be determined from data

Notes or active learning

Answer Key & Detailed Explanations

Answer Key

1: A	21: E	41: A	61: E	81: E	101: C
2: B	22: A	42: E	62: B	82: A	102: A
3: A	23: E	43: B	63: D	83: C	103: E
4: A	24: B	44: E	64: A	84: C	104: D
5: C	25: C	45: B	65: D	85: A	105: D
6: D	26: D	46: C	66: C	86: B	106: C
7: B	27: B	47: B	67: D	87: C	107: B
8: A	28: A	48: D	68: D	88: C	108: D
9: E	29: E	49: C	69: B	89: A	109: E
10: D	30: A	50: E	70: A	90: D	110: D
11: D	31: C	51: C	71: B	91: D	111: D
12: B	32: B	52: B	72: B	92: E	112: C
13: D	33: A	53: B	73: B	93: C	113: B
14: B	34: A	54: C	74: D	94: A	114: B
15: B	35: B	55: A	75: D	95: D	115: C
16: D	36: E	56: D	76: B	96: E	116: E
17: B	37: A	57: E	77: D	97: A	117: A
18: E	38: A	58: E	78: C	98: E	118: A
19: A	39: D	59: E	79: B	99: E	119: A
20: C	40: E	60: C	80: E	100: D	120: C

Practice Set 1: Questions 1–20

1. A is correct.

General formula for the equilibrium constant of a reaction:

$$a\text{A} + b\text{B} \leftrightarrow c\text{C} + d\text{D}$$

$$K_{\text{eq}} = ([\text{C}]^c \times [\text{D}]^d) / ([\text{A}]^a \times [\text{B}]^b)$$

For the reaction above:

$$K_{\text{eq}} = [\text{C}] / [\text{A}]^2 \times [\text{B}]^3$$

2. B is correct.

Two gases having the same temperature have the same average molecular kinetic energy.

Therefore,

$$(\tfrac{1}{2})m_A v_A^2 = (\tfrac{1}{2})m_B v_B^2$$

$$m_A v_A^2 = m_B v_B^2$$

$$(m_B / m_A) = (v_A / v_B)^2$$

$$(v_A / v_B) = 2$$

$$(m_B / m_A) = 2^2 = 4$$

The gases listed with a mass ratio of ~4 to 1 are iron (55.8 g/mol) and nitrogen (14 g/mol).

3. A is correct.

The slowest reaction would be the highest activation energy and the lowest temperature.

4. A is correct.

The rate law is calculated experimentally by comparing trials and determining how changes in the initial concentrations of the reactants affect the rate of the reaction.

$$\text{rate} = k[A]^x \cdot [B]^y$$

where k is the rate constant, and the exponents x and y are the partial reaction orders (i.e., determined experimentally). They are not equal to the stoichiometric coefficients.

To determine the order of reactant A, find two experiments where the concentrations of B are identical, and concentrations of A are different.

Use data from experiments 2 and 3 to calculate order of A:

$$\text{Rate}_3 / \text{Rate}_2 = ([A]_3 / [A]_2)^{\text{order of A}}$$

$$0.500 / 0.500 = (0.060 / 0.030)^{\text{order of A}}$$

$$1 = (2)^{\text{order of A}}$$

$$\text{Order of A} = 0$$

5. C is correct.

The balanced equation: $C_2H_6O + 3\ O_2 \rightarrow 2\ CO_2 + 3\ H_2O$

The rate of reaction/consumption is proportional to the coefficients.

The rate of carbon dioxide production is twice the rate of ethanol consumption:

$$2 \times 4.0\ \text{M s}^{-1} = 8.0\ \text{M s}^{-1}$$

6. D is correct.

General formula for the equilibrium constant of a reaction:

$$a\text{A} + b\text{B} \leftrightarrow c\text{C} + d\text{D}$$

$$K_{eq} = ([\text{C}]^c \times [\text{D}]^d) / ([\text{A}]^a \times [\text{B}]^b)$$

For equilibrium constant calculation, only include species in aqueous or gas phases.

$$K_{eq} = 1 / [\text{Cl}_2]$$

7. B is correct.

$$K_{eq} = [\text{products}] / [\text{reactants}]$$

If the K_{eq} is less than 1 (e.g., 6.3×10^{-14}), then the numerator (i.e., products) is smaller than the denominator (i.e., reactants), and fewer products have formed relative to the reactants.

If the reaction favors reactants compared to products, the equilibrium lies to the left.

8. A is correct.

The higher the energy of activation, the slower the reaction is. The height of this barrier is independent of the determination of spontaneous (ΔG is negative with products more stable than reactants) or nonspontaneous reactions (ΔG is positive with products less stable than reactants).

9. E is correct.

The main function of catalysts is lowering the reaction's activation energy (energy barrier), thus increasing its rate (k).

Temperature measures the average kinetic energy (i.e., $KE = \frac{1}{2}mv^2$) of the molecules.

Except for zero-order chemical reactions, increased concentration of reactants increases the probability that the reactants collide with sufficient energy and orientation to overcome the energy of activation barrier and proceed toward products.

10. D is correct.

Every reaction has activation energy: the amount of energy required by the reactant to start reacting.

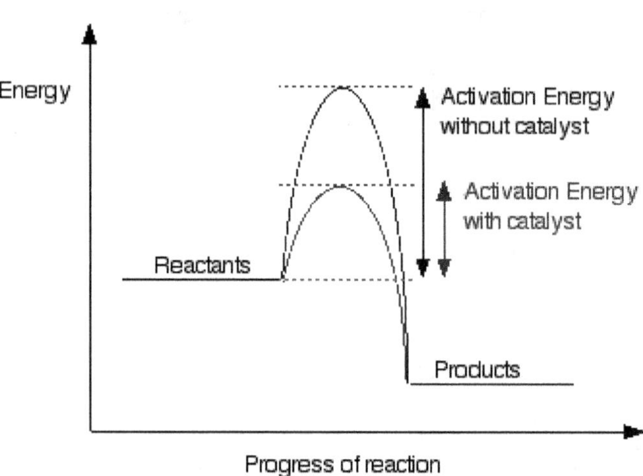

On the graph, activation energy can be estimated by calculating the distance between the initial energy level (i.e., reactant) and the peak (i.e., transition state). Activation energy is *not* the difference between energy levels of the initial and final state (reactant and product); ΔG.

A catalyst provides an alternative pathway for the reaction to proceed to product formation. It lowers the energy of activation (i.e., relative energy between reactants and transition state) and, therefore, speeds the reaction rate.

Catalysts do not affect the Gibbs free energy (ΔG: stability of products vs. reactants) or the enthalpy (ΔH: bond breaking in reactants or bond making in products).

11. D is correct.

The activation energy is the energy barrier overcome to transform reactants into products.

A reaction with higher activation energy (energy barrier) has a decreased rate (k).

12. B is correct.

The activation energy is the energy barrier overcome to transform reactants into products.

A reaction with lower activation energy (energy barrier) has an increased rate (k).

13. D is correct.

The activation energy is the energy barrier overcome to transform reactants into products.

A reaction with higher activation energy (energy barrier) has a decreased rate (k).

If the graphs are on the same scale, the height of the activation energy (R to the highest peak on the graph) is greatest in graph d.

14. B is correct.

If the graphs are on the same scale, the height of the activation energy (R to the highest peak on the graph) is the smallest in graph b.

The primary function of catalysts is lowering the reaction's activation energy (energy barrier), thus increasing its rate (k).

Although catalysts decrease the amount of energy required to reach the rate-limiting transition state, they do *not* decrease the relative energy of the products and reactants. Therefore, a catalyst does not affect ΔG.

A catalyst provides an alternative pathway for the reaction to proceed to product formation. It lowers the energy of activation (i.e., relative energy between reactants and transition state) and, therefore, speeds the reaction rate.

Catalysts do not affect the Gibbs free energy (ΔG: stability of products vs. reactants) or the enthalpy (ΔH: bond breaking in reactants or bond making in products).

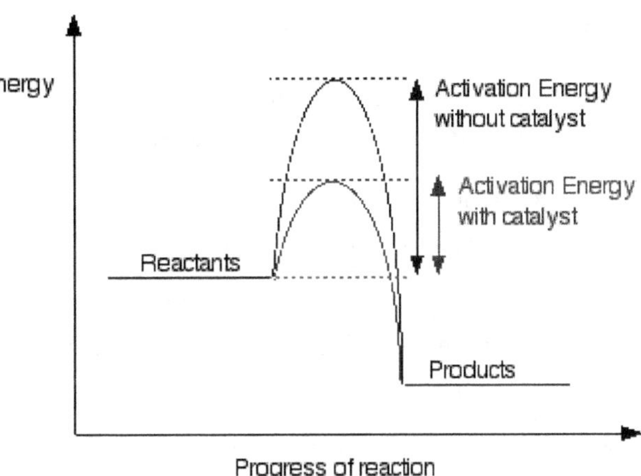

Catalysts do not affect ΔG (relative levels of R and P). Graph D shows an endergonic reaction with products less stable than reactants. The other graphs show an exergonic reaction with the same relative difference between the more stable and less stable reactants; the activation energy is different.

15. B is correct.

Achieving equilibrium is common for reactions. A reaction is at equilibrium when the forward reaction rate equals the reverse reaction rate.

Catalysts, by definition, are regenerated during a reaction and are not consumed by the reaction.

16. D is correct.

The molecules must collide with sufficient energy, frequency of collision, and proper orientation to overcome the activation energy barrier.

17. B is correct.

A catalyst lowers the energy of activation, which increases the rate of the reaction.

The primary function of catalysts is lowering the reaction's activation energy (energy barrier), thus increasing its rate (k).

Although catalysts decrease the amount of energy required to reach the rate-limiting transition state, they do *not* decrease the relative energy of the products and reactants.

Therefore, a catalyst does not affect ΔG.

18. E is correct.

When changing the reaction conditions, Le Châtelier's principle states that the position of equilibrium shifts to counteract the change.

If the concentration of reactants or products changes, the position of the equilibrium changes. Adding reactants or removing products shifts the equilibrium to the right.

According to Le Châtelier's Principle, adding or removing a solid at equilibrium does not affect equilibrium. However, when placed in an aqueous solution, solid NaOH dissociates into Na+ and OH–. The OH^- combines with H^+ to form water.

Reducing H^+ as a product drives the reaction to the right.

19. A is correct.

When changing the conditions of a reaction, Le Châtelier's principle states that the position of equilibrium shifts to counteract the change.

If the reaction temperature, pressure, or volume changes, the position of equilibrium changes.

Adding reactants or removing products shifts the equilibrium to the right.

Heat (i.e., energy related to temperature) is a reactant, and increasing its value drives the reaction toward product formation.

Decreasing the pressure on the reaction vessel drives the reaction toward reactants (i.e., to the left). The relative molar concentration of the reactants (i.e., $2 + 6$) is greater than the products (i.e., $4 + 3$).

The other choices listed drive the reaction toward reactants (i.e., to the left).

20. C is correct.

First, calculate the concentration/molarity of each reactant:

H_2 : 0.20 moles / 4.00 L = 0.05 M

X_2 : 0.20 moles / 4.00 L = 0.05 M

HX: 0.800 moles / 4.00L = 0.20 M

Use the concentrations to calculate Q:

$$Q = [HX]^2 / [H_2] \cdot [X_2]$$

$$Q = (0.20)^2 / (0.05 \times 0.05)$$

$$Q = 16$$

Because $Q < K_c$ (i.e., $16 < 24.4$), the reaction shifts to the right.

For Q to increase and match K_c, the numerator (i.e., $[HX]^2$), which is the product (right side) of the equation, needs to be increased.

Notes or active learning

Practice Set 2: Questions 21–40

21. E is correct.

General formula for the equilibrium constant of a reaction:

$$a\text{A} + b\text{B} \leftrightarrow c\text{C} + d\text{D}$$

$$K_{eq} = ([\text{C}]^c \times [\text{D}]^d) / ([\text{A}]^a \times [\text{B}]^b)$$

For equilibrium constant (K_c) calculation, only include species in aqueous or gas phases:

$$K_c = [CO_2]\,[H_2O]^2 / [CH_4]$$

22. A is correct.

The rate law is calculated by determining how changes in concentrations of the reactants affect the initial rate of the reaction.

$$\text{rate} = k[\text{A}]^x \cdot [\text{B}]^y$$

where k is the rate constant, and the exponents x and y are the partial reaction orders. They are not equal to the stoichiometric coefficients.

Whenever the fast (i.e., second) step follows the slow (i.e., first) step, the fast step is assumed to reach equilibrium, and the equilibrium concentrations are used for the rate law of the slow step.

23. E is correct.

The main function of catalysts is lowering the reaction's activation energy (energy barrier), thus increasing its rate (k).

Although catalysts decrease the amount of energy required to reach the rate-limiting transition state, they do *not* decrease the relative energy of the products and reactants.

Therefore, a catalyst does not affect ΔG.

continued...

A catalyst provides an alternative pathway for the reaction to proceed to product formation. It lowers the energy of activation (i.e., relative energy between reactants and transition state) and speeds the reaction rate.

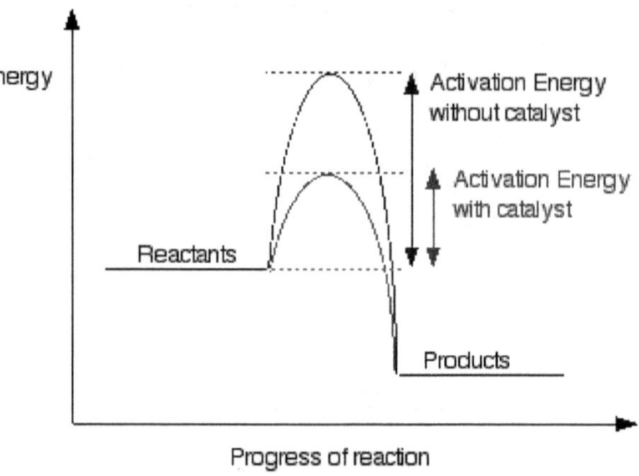

Catalysts do not affect the Gibbs free energy (ΔG: stability of products vs. reactants) or the enthalpy (ΔH: bond breaking in reactants or bond making in products).

24. B is correct.

Equilibrium reactions vary between conditions, so any proportion of product/reactant mass is possible. The question refers to the state *after* a reaction reaches equilibrium.

For a reaction that favors the formation of *products* ($K_{eq} > 1$), the amount of product would be more than reactants.

For a reaction that favors *reactants* ($K_{eq} < 1$), the amount of product would be less than reactants.

When $K_{eq} = 1$, the amounts of products and reactants would be equal.

25. C is correct.

The rate law is calculated experimentally by comparing trials and determining how changes in the initial concentrations of the reactants affect the rate of the reaction.

$$\text{rate} = k[A]^x \cdot [B]^y$$

where k is the rate constant, and the exponents x and y are the partial reaction orders (i.e., determined experimentally). They are not equal to the stoichiometric coefficients

Rate laws cannot be determined from the balanced equation (i.e., used to determine equilibrium) unless the reaction occurs in a single step.

From the data, when the concentration doubles, the rate quadruples.

Since 4 is 2^2, the rate law is second-order, therefore the rate = $k[H_2]^2$.

26. D is correct.

Lowering temperature shifts the equilibrium to the right because heat is a product (lowering the temperature is similar to removing the product). Increasing temperature (i.e., heating the system) has the opposite effect and shifts the equilibrium to the left (i.e., toward products).

Removing H_2 (i.e., reactant) shifts the equilibrium to the left.

Adding NH_3 (i.e., products) shifts the equilibrium to the left.

A catalyst lowers the energy of activation but does not affect the position of the equilibrium.

27. B is correct.

The order of the reaction has to be determined experimentally.

It is possible for a reaction's order to be identical to its coefficients.

For example, the coefficients correlate to the rate law in a single-step reaction or during the slow step of a multi-step reaction.

28. A is correct.

General formula for the equilibrium constant of a reaction:

$$a\text{A} + b\text{B} \leftrightarrow c\text{C} + d\text{D}$$

$$K_{eq} = ([\text{C}]^c \times [\text{D}]^d) / ([\text{A}]^a \times [\text{B}]^b)$$

For equilibrium constant calculation, only include species in aqueous or gas phases:

$$K_{eq} = 1 / [\text{CO}] \times [\text{H}_2]^2$$

29. E is correct.

Two peaks in this reaction indicate two energy-requiring steps with one intermediate (i.e., C) and each peak (i.e., B and D) as an activated complex (i.e., transition states).

The activated complex (i.e., transition state) is undergoing bond breaking/bond making events.

30. A is correct.

The activation energy for the slow step of a reaction is the distance from the starting material (or an intermediate) to the activated complex (i.e., transition state) with the absolute highest energy (i.e., highest point on the graph).

31. C is correct.

The activation energy for the slow step of a reverse reaction is the distance from an intermediate (or product) to the activated complex (i.e., transition state) with the absolute highest energy (i.e., highest point on the graph).

The slow step may not have the greatest magnitude for activation energy (e.g., E$\rightarrow$D on the graph).

32. B is correct.

The change in energy (or ΔH) is the difference between the energy of the reactants and products.

33. A is correct.

As the temperature (i.e., average kinetic energy) increases, the particles move faster (i.e., increased kinetic energy) and collide more frequently per unit time. This increases the reaction rate.

34. A is correct.

When changing the conditions of a reaction, Le Châtelier's principle states that equilibrium shifts to counteract the change.

If the reaction temperature, pressure, or volume is changed, the position of equilibrium changes.

According to Le Châtelier's principle, endothermic ($+\Delta H$) reactions increase the formation of products at higher temperatures.

Since both sides of the reaction have 2 gas molecules, a change in pressure does not affect equilibrium.

35. B is correct.

All chemical reactions eventually reach equilibrium, the state at which the reactants and products are present in concentrations that have no further tendency to change with time.

Therefore, the rate of production of each of the products (i.e., forward reaction) equals the rate of their consumption by the reverse reaction.

36. E is correct.

Chemical equilibrium refers to a dynamic process whereby the *rate* at which a reactant molecule is being transformed into a product is the same as the *rate* for a product molecule to be transformed into a reactant.

All chemical reactions eventually reach equilibrium, the state at which the reactants and products are present in concentrations with no further tendency to change with time.

37. A is correct.

Expression for equilibrium constant: $K = [H_2O]^2 \cdot [Cl_2]^2 / [HCl]^4 \cdot [O_2]$

Solve for $[Cl_2]$:

$$[Cl_2]^2 = (K \times [HCl]^4 \cdot [O_2]) / [H_2O]^2$$

$$[Cl_2]^2 = [46.0 \times (0.150)^4 \times 0.395] / (0.625)^2$$

$$[Cl_2]^2 = 0.0235$$

$$[Cl_2] = 0.153 \text{ M}$$

38. A is correct.

To determine the amount of each compound at equilibrium, consider the chemical reaction written in the form:

$$a\text{A} + b\text{B} \leftrightarrow c\text{C} + d\text{D}$$

The equilibrium constant (K_c) is defined as:

$$K_c = ([C]^c \times [D]^d) / ([A]^a \times [B]^b)$$

or $\qquad K_c = [\text{products}] / [\text{reactants}]$

If the K_{eq} is greater than 1, the numerator (i.e., products) is larger than the denominator (i.e., reactants), and more products have formed relative to the reactants.

If the reaction favors products compared to reactants, the equilibrium lies to the right.

39. D is correct.

Gases (as opposed to liquids and solids) are most sensitive to changes in pressure. There are three moles of hydrogen gas as a reactant and three moles of water vapor as a product.

Therefore, changes in pressure result in proportionate changes to the forward and reverse reactions.

40. E is correct.

When changing the conditions of a reaction, Le Châtelier's principle states that the position of equilibrium shifts to counteract the change.

If the reaction temperature, pressure, volume, or concentration changes, the position of the equilibrium changes.

CO is a reactant, and increasing its concentration shifts the equilibrium toward product formation (i.e., to the right).

Practice Set 3: Questions 41–60

41. A is correct.

General formula for the equilibrium constant of a reaction:

$$a\mathrm{A} + b\mathrm{B} \leftrightarrow c\mathrm{C} + d\mathrm{D}$$

$$K_{eq} = ([\mathrm{C}]^c \times [\mathrm{D}]^d) / ([\mathrm{A}]^a \times [\mathrm{B}]^b)$$

For equilibrium constant calculation, only include species in aqueous or gas phases:

$$K_{eq} = [\mathrm{NO}]^4 \times [\mathrm{H_2O}]^6 / [\mathrm{NH_3}]^4 \times [\mathrm{O_2}]^5$$

42. E is correct.

The high levels of CO_2 cause a person to hyperventilate to reduce the number of CO_2 (reactant).

Hyperventilating has the effect of driving the reaction toward products.

Removing products (i.e., HCO_3^- and H^+) or intermediates (H_2CO_3) drives the reaction toward products.

43. B is correct.

The units of the rate constants:

Zero-order reaction: $M\ sec^{-1}$

First-order reaction: sec^{-1}

Second-order reaction: $L\ mole^{-1}\ sec^{-1}$

44. E is correct.

A reaction proceeds when the reactant(s) have sufficient energy to overcome activation energy and proceed to products.

Increasing the temperature increases the molecule's kinetic energy, increases the frequency of collision, and increases the probability that the reactants will overcome the barrier (energy of activation) to form products.

45. B is correct.

Low activation energy increases the rate because the reaction's required energy is lower.

High temperature increases the rate because faster-moving molecules have a greater probability of collision, facilitating the reaction.

Combined, lower activation energy and a higher temperature results in the highest relative rate for the reaction.

46. C is correct.

The rate law is calculated by comparing trials and determining how changes in the initial concentrations of the reactants affect the rate of the reaction.

$$\text{rate} = k[\text{A}]^x \cdot [\text{B}]^y$$

where k is the rate constant, and the exponents x and y are the partial reaction orders (i.e., determined experimentally); they are not equal to the stoichiometric coefficients.

Start by identifying two reactions where XO concentration is constant and O_2 is different: experiments 1 and 2.

When the concentration of O_2 is doubled, the rate is doubled. This indicates that the order of the reaction for O_2 is 1.

Now, determine the order for *X*O.

Find 2 reactions where the concentration of O_2 is constant, and *X*O is different: experiments 2 and 3.

When the concentration of *X*O is tripled, the rate is multiplied by a factor of 9.

$3^2 = 9$, which means the order of the reaction with respect to *X*O is 2.

Therefore, the expression of rate law is:

$$\text{rate} = k[X\text{O}]^2 \cdot [\text{O}_2]$$

47. B is correct.

General formula for the equilibrium constant of a reaction:

$$a\text{A} + b\text{B} \leftrightarrow c\text{C} + d\text{D}$$

$$K_{eq} = ([\text{C}]^c \times [\text{D}]^d) / ([\text{A}]^a \times [\text{B}]^b)$$

For equilibrium constant calculation, only include species in aqueous or gas phases.

$$K_{eq} = 1 / [\text{CO}_2]$$

48. D is correct.

$$K_{eq} = [\text{products}] / [\text{reactants}]$$

If the numerator (i.e., products) is smaller than the denominator (i.e., reactants), and fewer products have formed relative to the reactants. Therefore, K_{eq} is less than 1.

49. C is correct.

The activation energy is the energy barrier for a reaction to proceed. The forward reaction is measured from the reactants to the highest energy level in the reaction.

A reaction mechanism with low energy of activation proceeds faster than a reaction with high energy of activation.

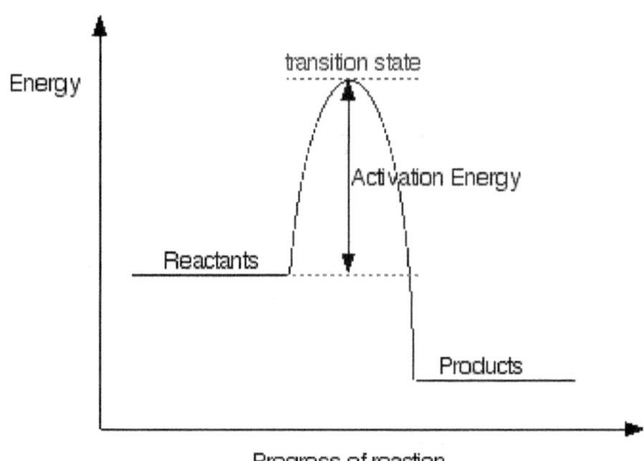

50. E is correct.

As the average kinetic energy (i.e., $KE = \tfrac{1}{2}mv^2$) increases, the particles move faster and collide more frequently per unit time and possess greater energy when they collide. This increases the reaction rate. Hence the reaction rate of most reactions increases with increasing temperature.

For reversible reactions, it is common for the rate law to depend on the concentration of the products. The overall rate is negative (except for autocatalytic reactions).

As the concentration of products decreases, their collision frequency decreases, and the reverse reaction rate decreases; therefore, the overall rate of the reaction increases.

The primary function of catalysts is lowering the reaction's activation energy (energy barrier), thus increasing its rate (k).

51. C is correct.

General formula for the equilibrium constant of a reaction:

$$a\text{A} + b\text{B} \leftrightarrow c\text{C} + d\text{D}$$

$$K_{eq} = ([\text{C}]^c \times [\text{D}]^d) / ([\text{A}]^a \times [\text{B}]^b)$$

For the ionization equilibrium (K_i) constant calculation, only include species in aqueous or gas phases (which, in this case, is all):

$$K_i = [\text{H}^+]\cdot[\text{HS}^-] / [\text{H}_2\text{S}]$$

52. B is correct.

As the temperature (i.e., average kinetic energy) increases, the particles move faster and collide more frequently per unit time and possess greater energy when they collide. This increases the reaction rate. Hence the reaction rate of most reactions increases with increasing temperature.

53. B is correct.

Reactions are spontaneous when Gibbs free energy (ΔG) is negative, and the reaction is described as exergonic; the products are more stable than the reactants.

Reactions are nonspontaneous when Gibbs free energy (ΔG) is positive, and the reaction is described as endergonic; the products are less stable than the reactants.

54. C is correct.

When changing the conditions of a reaction, Le Châtelier's principle states that equilibrium shifts to counteract the change. If the reaction temperature, pressure, or volume is changed, the position of equilibrium changes.

There are 2 moles on each side, so changes in pressure (or volume) do not shift the equilibrium.

55. A is correct.

All of the choices are possible events of chemical reactions, but they do not need to happen for a reaction to occur, except that reactant particles must collide (i.e., make contact) with each other.

56. D is correct.

The enthalpy or heat of reaction (ΔH) is not a function of temperature.

Since the reaction is exothermic, Le Châtelier's principle states that increasing the temperature decreases the forward reaction.

57. E is correct.

When changing the reaction conditions, Le Châtelier's principle states that the position of equilibrium shifts to counteract the change. If the reaction's concentration changes, the position of the equilibrium changes.

According to Le Châtelier's Principle, adding or removing a solid at equilibrium does not affect equilibrium. However, adding solid $KC_2H_3O_2$ is equivalent to adding K^+ and $C_2H_3O_2^-$ (i.e., product) as it dissociates in an aqueous solution. The increased concentration of products shifts the equilibrium toward reactants (i.e., to the left).

Increasing the pH decreases the $[H^+]$ and would shift the reaction to products (i.e., right).

58. E is correct.

When changing the conditions of a reaction, Le Châtelier's principle states that the position of equilibrium shifts to counteract the change. If the reaction temperature, pressure, or volume is changed, the position of equilibrium changes.

The $K_{eq} = 2.8 \times 10^{-21}$ which indicates a low concentration of products compared to reactants:

$$K_{eq} = [\text{products}] / [\text{reactants}]$$

59. E is correct.

Chemical equilibrium refers to a dynamic process whereby the rate at which a reactant molecule is being transformed into a product is the same as the rate for a product molecule to be transformed into a reactant. The rate of the forward reaction is equal to the rate of the reverse reaction.

60. C is correct.

All chemical reactions eventually reach equilibrium, the state at which the reactants and products are present in concentrations with no further tendency to change with time.

Catalysts speed up reactions and increase the rate at which equilibrium is reached.

However, they never alter the thermodynamics of a reaction and therefore do not change free energy ΔG.

Catalysts do not alter the equilibrium constant.

Notes for active learning

Practice Set 4: Questions 61–80

61. E is correct.

To determine the order with respect to W, compare the data for trials 2 and 4.

The concentrations of X and Y do not change when comparing trials 2 and 4, but the concentration of W changes from 0.015 to 0.03, which corresponds to an increase by a factor of 2.

However, the rate did not increase (0.08 remains); therefore, the order with respect to W is 0.

The order for X is determined by comparing the data for trials 1 and 3.

The concentrations for W and Z are constant, and the concentration for X increases by a factor of 3 (from 0.05 to 0.15).

The rate of the reaction increased by a factor of 9 (0.04 to 0.36).

Since $3^2 = 9$, the reaction order with respect to X is two.

The order with respect to Y is found by comparing the data from trials 1 and 5. [W] and [X] do not change, but [Y] goes up by a factor of 4.

The rate from trials 1 to 5 goes up by a factor of 2.

Since $4^{1/2} = 2$, the reaction order with respect to Y is ½.

The overall order is found by the sum of the orders for X, Y, and Z:

$$0 + 2 + \tfrac{1}{2} = 2\tfrac{1}{2}$$

62. B is correct.

Since the order with respect to W is zero, the rate of formation of Z does not depend on the concentration of W.

63. D is correct.

The rate law is calculated by comparing trials and determining how changes in the initial concentrations of the reactants affect the rate of the reaction.

$$\text{rate} = k[A]^{x} \cdot [B]^{y}$$

where k is the rate constant, and the exponents x and y are the partial reaction orders (i.e., determined experimentally). They are not equal to the stoichiometric coefficients.

Use the partial orders determined in question **61** to express the rate law:

$$\text{rate} = k[W]^{0}[X]^{2}[Z]^{\frac{1}{2}}$$

Substitute the values for trial #1:

$$0.04 = k\,[0.01]^{0}[0.05]^{2}[0.04]^{\frac{1}{2}}$$

$$0.04 = (k) \cdot (1) \cdot (2.5 \times 10^{-3}) \cdot (2 \times 10^{-1})$$

$$k = 80$$

64. A is correct.

When changing the conditions of a reaction, Le Châtelier's principle states that the position of the equilibrium shifts to counteract the change.

If the reaction temperature, pressure, or volume changes, the position of the equilibrium changes.

Decreasing the temperature of the system favors the production of more heat. It shifts the reaction towards the exothermic side.

Because $\Delta H > 0$ (i.e., heat is a reactant), the reaction is endothermic.

Therefore, decreasing the temperature shifts the equilibrium toward the reactants.

65. D is correct.

General formula for the equilibrium constant of a reaction:

$$aA + bB \leftrightarrow cC + dD$$

$$K_{eq} = ([C]^{c} \times [D]^{d}) / ([A]^{a} \times [B]^{b})$$

For the equilibrium constant (K_{i}), only include species in aqueous phases:

$$K_{i} = [H^{+}] \times [H_2PO_4^{-}] / [H_3PO_4]$$

66. C is correct.

Catalysts speed up reactions and increase the rate at which equilibrium is reached. However, catalysts do not alter the thermodynamics of a reaction and therefore do not change free energy ΔG.

A catalyst provides an alternative pathway for the reaction to proceed to product formation. It lowers the energy of activation (i.e., relative energy of reactants and transition state) and speeds the reaction rate.

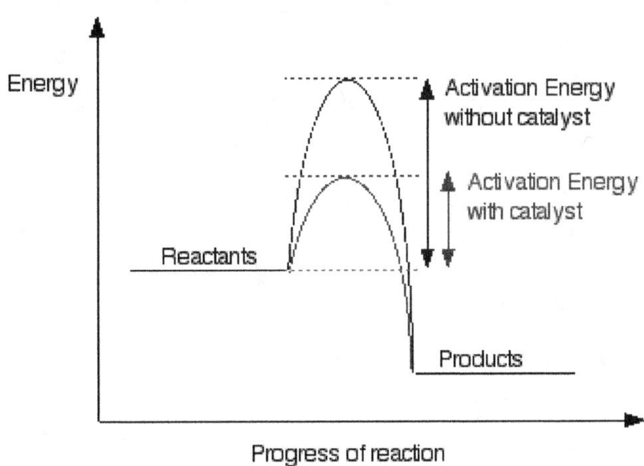

Catalysts do not affect the Gibbs free energy (ΔG: stability of products vs. reactants) or the enthalpy (ΔH: bond breaking in reactants or bond making in products).

67. D is correct.

General formula for the equilibrium constant of a reaction:

$$a\text{A} + b\text{B} \leftrightarrow c\text{C} + d\text{D}$$

$$K_{eq} = ([\text{C}]^c \times [\text{D}]^d) / ([\text{A}]^a \times [\text{B}]^b)$$

For the ionization equilibrium constant (K_i) calculation, only include species in aqueous or gas phases (which, in this case, is):

$$K_i = [\text{H}^+] \cdot [\text{HSO}_3^-] / [\text{H}_2\text{SO}_3]$$

68. D is correct.

$$K_{eq} = [\text{products}] / [\text{reactants}]$$

If the K_{eq} is greater than 1, then the numerator (i.e., products) is greater than the denominator (i.e., reactants), and more products have formed relative to the reactants.

69. B is correct.

The activation energy is the energy barrier for a reaction to proceed.

The forward reaction is measured from the energy of the reactants to the highest energy level in the reaction.

70. A is correct.

Lowering the temperature decreases the reaction rate because the molecules involved in the reaction move and collide more slowly, so the reaction occurs at a slower speed.

Increasing the concentration of reactants due to the increased probability that the reactants collide with sufficient energy and orientation to overcome the energy of the activation barrier and proceed toward products. The primary function of catalysts is lowering the reaction's activation energy (energy barrier), thus increasing its rate (k).

71. B is correct.

The activation energy is the energy barrier that must be overcome to transform reactant(s) into products. The molecules must collide with the proper orientation and energy (i.e., kinetic energy is the average temperature) to overcome the energy of the activation barrier.

72. B is correct.

The activation energy is the energy barrier that must be overcome to transform reactant(s) into products.

The molecules must collide with the proper orientation and energy (i.e., kinetic energy is the average temperature) to overcome the energy of the activation barrier.

Solids describe molecules with limited relative motion.

Liquids are molecules with more motion than solids, while gases exhibit the most motion.

73. B is correct.

The rate law is calculated experimentally by comparing trials and determining how changes in the initial concentrations of the reactants affect the rate of the reaction.

$$\text{rate} = k[A]^x \cdot [B]^y$$

where k is the rate constant, and the exponents x and y are the partial reaction orders (i.e., determined experimentally). They are not equal to the stoichiometric coefficients.

Comparing Trials 1 and 3, [A] increased by a factor of 3, as did the reaction rate; thus, the reaction is first order with respect to A.

Comparing Trials 1 and 2, [B] increased by a factor of 4 and the reaction rate increased by a factor of $16 = 4^2$.

Thus, the reaction is second order with respect to B. Therefore, the rate $= k[A] \cdot [B]^2$

74. D is correct.

To overcome the *activation energy* barrier, the molecules must collide with sufficient energy, frequency, and proper orientation.

75. D is correct.

When changing the conditions of a reaction, Le Châtelier's principle states that equilibrium shifts to counteract the change.

If the reaction temperature, pressure, or volume is changed, the position of equilibrium changes.

Removing NH_3 and adding N_2 shifts the equilibrium to the right.

Removing N_2 or adding NH_3 shifts the equilibrium to the left.

76. B is correct.

Increasing $[HC_2H_3O_2]$ and $[H^+]$ affects the equilibrium because both species are part of the equilibrium equation.

$$K_{eq} = [\text{products}] / [\text{reactants}]$$

$$K_{eq} = [C_2H_3O_2^-] / [HC_2H_3O_2] \cdot [H^+]$$

Adding solid $NaC_2H_3O_2$: in aqueous solutions, $NaC_2H_3O_2$ dissociates into Na^+ (*aq*) and $C_2H_3O_2^-$ (*aq*).

Therefore, the overall concentration of $C_2H_3O_2^-$ (*aq*) increases, affecting the equilibrium.

Adding solid $NaNO_3$: this is a salt and does not react with reactant or product molecules; therefore, it will not affect the equilibrium.

Adding solid NaOH: the reactant is an acid because it dissociates into hydrogen ions and anions in aqueous solutions. It reacts with bases (e.g., NaOH) to create water and salt.

Thus, NaOH reduces the concentration of the reactant, which affects the equilibrium.

Increasing $[HC_2H_3O_2]$ and $[H^+]$ increases reactants and drives the reaction toward product formation.

77. D is correct.

Calculate the value of the equilibrium expression: hydrogen iodide concentration decreases for equilibrium to be reached.

$$K = [\text{products}] / [\text{reactants}]$$

$$K = [HI]^2 / [H_2] \cdot [I_2]$$

To determine the state of a reaction, calculate its reaction quotient (Q).

Q has the same calculation method as equilibrium constant (K); the difference is that K has to be calculated at the point of equilibrium, whereas Q can be calculated at any time.

If $Q > K$, reaction shifts towards reactants (more reactants, fewer products)

If $Q = K$, reaction is at equilibrium

If $Q < K$, reaction shifts towards products (more products, less reactants)

$$Q = [HI]^2 / [H_2] \cdot [I_2]$$

$$Q = (3)^2 / 0.4 \times 0.6$$

$$Q = 37.5$$

Because $Q > K$ (i.e., (37.5 > 35), the reaction shifts toward reactants (larger denominator), and the amount of product (HI) decreases.

78. C is correct.

Decreasing the concentration of reactants (or decreasing pressure) decreases the reaction rate because there is a decreased probability that any two reactant molecules collide with sufficient energy to overcome the energy of activation and form products.

79. B is correct.

Chemical equilibrium expression is dependent on the stoichiometry of the reaction, more specifically, the coefficients of each species involved in the reaction.

The mechanism refers to the pathway of product formation (e.g., S_N1 or S_N2) but does not affect the relative energies of the reactants and products (i.e., a position for the equilibrium).

The rate refers to the time needed to achieve equilibrium but does not affect the equilibrium position.

80. E is correct.

$$K = [products] / [reactants]$$

$$K = [CH_3OH] / [H_2O]^2 \cdot [CO]$$

None of the stated changes tend to decrease the *magnitude* of the equilibrium constant K.

For exothermic reactions ($\Delta H < 0$), decreasing the temperature increases the magnitude of K.

Changes in volume, pressure, or concentration do not affect the magnitude of K.

Practice Set 5: Questions 81–100

81. E is correct.

When changing the reaction conditions, Le Châtelier's principle states that the position of equilibrium shifts to counteract the change.

If the reaction's concentration changes, the position of the equilibrium changes.

Adding reactants or removing products shifts the equilibrium to the right (i.e., product formation).

82. A is correct.

When two or more reactions are combined to create a new reaction, the K_c of the resulting reaction is a product of K_c values of the individual reactions.

When a reaction is reversed, the new $K_c = 1$ / old K_c.

Start by analyzing the reactions provided the problem:

$$(1)\quad 2\,NO + Cl_2 \leftrightarrow 2\,NOCl \qquad\qquad K_c = 3.2 \times 10^3$$

$$(2)\quad 2\,NO_2 \leftrightarrow 2\,NO + O_2 \qquad\qquad K_c = 15.5$$

$$(3)\quad NOCl + \tfrac{1}{2}\,O_2 \leftrightarrow NO_2 + \tfrac{1}{2}\,Cl_2 \qquad\qquad K_c = ?$$

Reaction 3 can be created by combining reactions 1 and 2.

For reaction 1, notice that on reaction 3, NOCl is located on the left.

To match reaction 3, reverse reaction 1:

$$2\,NOCl \leftrightarrow 2\,NO + Cl_2$$

The K_c of this reversed reaction is:

$$K_c = 1 / (3.2 \times 10^{-3}) = 312.5$$

Reaction 2 needs to be reversed to match the position of NO in reaction 3:

$$2\,NO + O_2 \leftrightarrow 2\,NO_2$$

The new K_c will be:

$$K_c = 1 / 15.5 = 0.0645$$

continued…

Since K of the resulting reaction is the product of K_c from both reactions, add those reactions:

$$2\ NOCl \leftrightarrow 2\ NO + Cl_2$$

$$2\ NO + O_2 \leftrightarrow 2\ NO_2$$

$$2\ NOCl + O_2 \leftrightarrow 2\ NO_2 + Cl_2$$

$$K_c = 312.5 \times 0.064$$

$$K_c = 20.16$$

Divide the equation by 2:

$$NOCl + \tfrac{1}{2}\ O_2 \leftrightarrow NO_2 + \tfrac{1}{2}\ Cl_2$$

The new K_c is the square root of initial K_c:

$$K_c = \sqrt{20.16} = 4.49$$

83. C is correct.

Increased pressure or increased concentration of reactants increases the probability that the reactants collide with sufficient energy and orientation to overcome the *energy of activation* barrier and proceed toward products.

84. C is correct.

The reaction rate is the speed at which reactants are consumed, or products are formed.

The rate of a chemical reaction at a constant temperature depends only on the concentrations of the substances that influence the rate. The reactants influence the rate of reaction, but occasionally products can influence the reaction rate.

85. A is correct.

General formula for the equilibrium constant of a reaction:

$$aA + bB \leftrightarrow cC + dD$$

$$K_{eq} = ([C]^c \times [D]^d) / ([A]^a \times [B]^b)$$

For the reaction above:

$$K_{eq} = [B] \times [C]^3 / [A]^2$$

86. B is correct.

The rate law is calculated by comparing trials and determining how changes in the initial concentrations of the reactants affect the rate of the reaction.

$$\text{rate} = k[A]^x \cdot [B]^y$$

where k is the rate constant, and the exponents x and y are the partial reaction orders (i.e., determined experimentally). They are not equal to the stoichiometric coefficients.

87. C is correct.

Rates of formation/consumption of species in a reaction are proportional to their coefficients.

If the rate of a species is known, the rate of other species can be calculated using simple proportions:

$$\text{coefficient}_{NOBr} \,/\, \text{coefficient}_{Br2} = \text{rate}_{formation\ NOBr} \,/\, \text{rate}_{consumption\ Br2}$$

$$\text{rate}_{consumption\ Br2} = \text{rate}_{formation\ NOBr} \,/\, (\text{coefficient}_{NOBr} \,/\, \text{coefficient}_{Br2})$$

$$\text{rate}_{consumption\ Br2} = 4.50 \times 10^{-4}\ \text{mol L}^{-1}\ \text{s}^{-1} \,/\, (2\,/\,1)$$

$$\text{rate}_{consumption\ Br2} = 2.25 \times 10^{-4}\ \text{mol L}^{-1}\ \text{s}^{-1}$$

88. C is correct.

The activation energy is the energy barrier to transform reactants into products.

A reaction with lower activation energy (energy barrier) has an increased rate (k).

89. A is correct.

General formula for the equilibrium constant of a reaction:

$$aA + bB \leftrightarrow cC + dD$$

$$K_{eq} = ([C]^c \times [D]^d) \,/\, ([A]^a \times [B]^b)$$

For the reaction above:

$$K_{eq} = [C]^2 \cdot [D] \,/\, [A] \cdot [B]^2$$

90. D is correct.

When changing the conditions of a reaction, Le Châtelier's principle states that the position of the equilibrium shifts to counteract the change.

If the reaction's water concentration (i.e., product) changes, equilibrium changes toward reactants.

91. D is correct.

For the general equation:

$$a\text{A} + b\text{B} \leftrightarrow c\text{C} + d\text{D}$$

The equilibrium constant is defined as:

$$K_{eq} = ([\text{C}]^c \times [\text{D}]^d) / ([\text{A}]^a \times [\text{B}]^b)$$

or

$$K_{eq} = [\text{products}] / [\text{reactants}]$$

If K_{eq} is less than 1 (e.g., 4.3×10^{-17}), then the numerator (i.e., products) is smaller than the denominator (i.e., reactants), and fewer products have formed relative to the reactants.

If the reaction favors reactants compared to products, the equilibrium lies to the left.

92. E is correct.

The primary function of catalysts is lowering the reaction's activation energy (energy barrier), thus increasing its rate (k). The activation energy is the energy barrier for a reaction to proceed. The forward reaction is measured from the energy of the reactants to the highest energy level in the reaction.

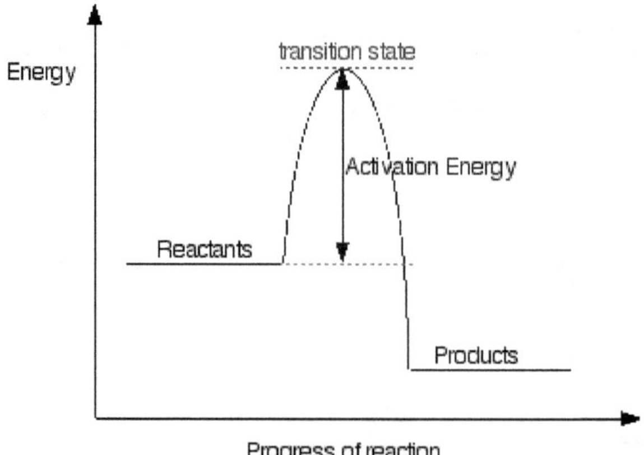

The transition state signifies "bond making" and "bond making" events in transforming the reactants into products.

93. C is correct.

Use the equilibrium concentrations to calculate K_c:

$$K_c = [\text{N}_2] \cdot [\text{H}_2]^3 / [\text{NH}_3]^2$$

$$K_c = 0.04 \times (0.12)^3 / (0.4)^2$$

$$K_c = 4.3 \times 10^{-4}$$

94. A is correct.

Increasing the concentration of reactants (or increasing pressure) increases the reaction rate because there is an increased probability that any two reactant molecules collide with sufficient energy to overcome the energy of activation and form products.

95. D is correct.

When changing the conditions of a reaction, Le Châtelier's principle states that equilibrium will shift to counteract the change.

If the reaction temperature, pressure, or volume is changed, the position of the equilibrium changes.

Add the molar coefficients in the reaction (i.e., 3 for the reactants and 1 for the product).

In general, decreasing the pressure favors the reaction's side with a higher molar coefficient sum (i.e., toward the reactants in this example).

Decreasing the volume favors the reaction's side with a smaller molar coefficient sum (i.e., products in this example).

Therefore, it would increase the yield of methanol.

Decreasing the temperature of the system favors the production of more heat. It shifts the reaction towards the exothermic side.

Because $\Delta H < 0$, the reaction is exothermic in the forward direction (towards product). Therefore, decreasing the temperature increases the yield of methanol.

96. E is correct.

Decreasing the pH: increases the concentration of H^+, which shifts the equilibrium to the left (i.e., increases the chlorine concentration)

Adding HCl: increases the concentration of H^+, which shifts the equilibrium to the left.

Adding HClO: increases the concentration of HClO, which shifts the equilibrium to the left.

Adding NaClO: increases the concentration of ClO^-, which shifts the equilibrium to the left.

97. A is correct.

The diffusion rate is proportional to the speed of the gas.

Since lighter gases travel faster than heavier gases, H_2 diffuses faster than O_2.

At the same temperature, two gases have the same average molecular kinetic energy:

$$\tfrac{1}{2}m_H v_H^2 = \tfrac{1}{2}m_O v_O^2$$

$$(v_H / v_O)^2 = m_O / m_H$$

Since,

$$m_O / m_H = 16$$

$$(v_H / v_O)^2 = 16$$

$$v_H / v_O = 4{:}1 \text{ ratio}$$

98. E is correct.

The primary function of catalysts is lowering the reaction's activation energy (energy barrier), thus increasing its rate (k).

Although catalysts decrease the amount of energy required to reach the rate-limiting transition state, they do *not* decrease the relative energy of the products and reactants.

Therefore, a catalyst does not affect ΔG.

Catalysts speed up reactions and increase the rate at which equilibrium is reached.

However, catalysts never alter the thermodynamics of a reaction and therefore do not change free energy ΔG.

99. E is correct.

Balanced reaction:

$$3\,A + 2\,B \rightarrow 4\,C$$

Moles of A used in the reaction = 1.4 moles − 0.9 moles = 0.5 moles

Moles of C formed = (coefficient C / coefficient A) × moles of A used

Moles of C formed = (4 / 3) × 0.5 moles = 0.67 moles

100. D is correct.

When two or more reactions are combined to create a new reaction, the K_c of the resulting reaction is a product of K_c values of the individual reactions.

When a reaction is reversed, new $K_c = 1$ / old K_c.

$$PCl_3 + Cl_2 \leftrightarrow PCl_5 \qquad K_c = K_{c1}$$

$$2\ NO + Cl_2 \leftrightarrow 2\ NOCl \qquad K_c = K_{c2}$$

Reverse the first reaction:

$$PCl_5 \leftrightarrow PCl_3 + Cl_2 \qquad K_c = 1 / K_{c1}$$

Add the two reactions to create a third reaction:

$$PCl_5 \leftrightarrow PCl_3 + Cl_2 \qquad K_c = 1 / K_{c1}$$

$$2\ NO + Cl_2 \leftrightarrow 2\ NOCl \qquad K_c = K_{c2}$$

$$PCl_5 + 2\ NO \leftrightarrow PCl_3 + 2\ NOCl$$

$$K = [(1 / K_1) \times K_2] = K_2 / K_1$$

Notes for active learning

Practice Set 6: Questions 101–120

101. C is correct.

Mass action expression is related to the equilibrium constant expression.

General formula for the equilibrium constant of a reaction:

$$a\mathrm{A} + b\mathrm{B} \leftrightarrow c\mathrm{C} + d\mathrm{D}$$

$$K_{eq} = ([\mathrm{C}]^c \times [\mathrm{D}]^d) / ([\mathrm{A}]^a \times [\mathrm{B}]^b)$$

To determine the mass action expression, only include species in aqueous or gas phases:

$$K_c = [\mathrm{CrCl_3}]^4 / [\mathrm{CCl_4}]^3$$

102. A is correct.

When changing the conditions of a reaction, Le Châtelier's principle states that the position of the equilibrium shifts to counteract the change.

If the reaction's concentration is changed, the position of equilibrium changes.

Adding KNO_2 increases the amount of K+ and NO_2^- (product).

Therefore, the reaction shifts toward the reactants (i.e., to the left).

103. E is correct.

Rate expressions are determined experimentally from the rate data for a reaction.

The calculation is not similar to that for the equilibrium constant.

Since rate data are not given for this reaction, the rate expression cannot be determined.

104. D is correct.

The rate law is calculated experimentally by comparing trials and determining how changes in the initial concentrations of the reactants affect the rate of the reaction.

$$\text{rate} = k[\mathrm{A}]^x \cdot [\mathrm{B}]^y$$

where k is the rate constant, and the exponents x and y are the partial reaction orders (i.e., determined experimentally). They are not equal to the stoichiometric coefficients

To write a complete rate law, the order of each reactant is required.

Determine the order of reactant D by using 2 experiments where the concentrations of E are identical, and concentrations of D are different: experiments 1 and 2.

Use data from these experiments to calculate order of D:

$$\text{Rate}_2 / \text{Rate}_1 = ([D]_2 / [D]_1)^{\text{order of D}}$$

$$(0.000500) / (0.000250) = (0.200 / 0.100)^{\text{order of D}}$$

$$2 = (2)^{\text{order of D}}$$

Order of D = 1

Use the same procedure to calculate order of E.

Data from experiments 1 and 3:

$$\text{Rate}_3 / \text{Rate}_1 = ([E]_3 / [E]_1)^{\text{order of E}}$$

$$(0.001000) / (0.000250) = (0.500 / 0.250)^{\text{order of E}}$$

$$4 = (2)^{\text{order of E}}$$

Order of E = 2

$$\text{Rate} = k[D]{\cdot}[E]^2$$

105. D is correct.

The rate of formation/consumption of species in a reaction is proportional to their coefficients.

If the rate of a species is known, the rate of other species can be calculated using simple proportions.

Because the formation rate of a species is already known, there is no need to calculate it.

(The extra information on the problem is just there to distract test-takers).

$$\text{rate}_{\text{consumption H2}} / \text{rate}_{\text{formation NO2}} = \text{coefficient}_{\text{H2}} / \text{coefficient}_{\text{NO2}}$$

$$\text{rate}_{\text{consumption H2}} = (\text{coefficient}_{\text{H2}} / \text{coefficient}_{\text{NO2}}) \times \text{rate}_{\text{formation NO2}}$$

$$\text{rate}_{\text{consumption H2}} = (4 / 2) \times 2.8 \times 10^{-4} \text{ M/min}$$

$$\text{rate}_{\text{consumption H2}} = 5.6 \times 10^{-4} \text{ M/min}$$

106. C is correct.

General formula for the equilibrium constant of a reaction:

$$a\text{A} + b\text{B} \leftrightarrow c\text{C} + d\text{D}$$

$$K_{eq} = ([\text{C}]^c \times [\text{D}]^d) / ([\text{A}]^a \times [\text{B}]^b)$$

For equilibrium constant calculation, only include species in aqueous or gas phases (which, in this problem, is all of them):

$$K_{eq} = [CO_2]^2 / [CO]^2 \times [O_2]$$

107. B is correct.

Reactions proceed faster at higher temperatures.

Catalysts lower the activation energy (i.e., energy barrier) for the reaction, and the reaction rate increases.

108. D is correct.

For the general equation:

$$a\text{A} + b\text{B} \leftrightarrow c\text{C} + d\text{D}$$

The equilibrium constant is defined as:

$$K_{eq} = ([\text{C}]^c \times [\text{D}]^d) / ([\text{A}]^a \times [\text{B}]^b)$$

or

$$K_{eq} = [\text{products}] / [\text{reactants}]$$

If K_{eq} is less than 1 (e.g., 6.3×10^{-14}), then the numerator (i.e., products) is smaller than the denominator (i.e., reactants), and fewer products have formed relative to the reactants.

If the reaction favors reactants compared to products, the equilibrium lies to the left.

109. E is correct.

Typically, the forward reaction's constant is denoted as k_1 and the reverse is k_{-1}.

The overall constant of step 1 would be k_1 / k_{-1}.

110. D is correct.

A catalyst lowers the activation energy (E_a) of a reaction.

The primary function of catalysts is lowering the reaction's activation energy (energy barrier), thus increasing its rate (k).

Although catalysts decrease the amount of energy required to reach the rate-limiting transition state, they do *not* decrease the relative energy of the products and reactants.

A catalyst provides an *alternative pathway* for the reaction to proceed to product formation. It lowers the activation energy (i.e., relative energy between reactants and transition state) and speeds the reaction rate.

Catalysts do *not* affect the Gibbs free energy (ΔG: stability of products vs. reactants) or the enthalpy (ΔH: bond breaking in reactants or bond making in products).

111. D is correct.

When changing the reaction conditions, Le Châtelier's principle states that the position of equilibrium shifts to counteract the change.

If the reaction's concentration changes, the position of equilibrium changes.

Adding reactants or removing products shifts the equilibrium to the right.

112. C is correct.

Increased pressure or increased concentration of reactants increases the probability of colliding with sufficient energy and orientation to overcome the energy of the activation barrier and proceed toward products.

113. B is correct.

Increasing the concentration of reactants (or increasing pressure) increases the reaction rate because there is an increased probability that any two reactant molecules collide with sufficient energy to overcome the energy of activation and form products.

114. B is correct.

The number of collisions does not have to be large for a reaction.

However, a minimum threshold of collisions needs to be reached before a reaction can proceed.

Reactions tend to happen more readily at higher temperatures because higher temperatures result in more collisions.

Some reactions only break bonds but do not form new bonds (e.g., radical cleavage of chlorine in the ozone decomposition process). Some reactions form bonds and do not break bonds (e.g., $Na^+ + Cl^- \rightarrow NaCl$).

It is not necessary that all chemical bonds in the reactants break for a reaction to occur.

115. C is correct.

Le Châtelier's principle states that *equilibrium shifts* to counteract the change when changing the reaction conditions.

If the reaction temperature, pressure, or volume is changed, the position of equilibrium changes.

A catalyst lowers the reaction's activation energy (energy barrier) and therefore increases the rate (k) of the reaction.

A catalyst does not affect the relative energy of the reactants and products and does not affect the equilibrium.

116. E is correct.

All chemical reactions eventually reach equilibrium, the state at which the reactants and products are present in concentrations with no further tendency to change with time.

When the concentration in the reactant side of a reaction is reduced, some product transforms (i.e., the reaction is reversed) into reactant to compensate for the reduction of reactants. This results in a decrease in the concentration from the product side of the reaction.

117. A is correct.

Activated complexes (i.e., transition state for molecules undergoing bond breaking and bond making) decompose rapidly have specific geometries that are extremely reactive.

Activated complexes (as opposed to intermediates) cannot be isolated.

118. A is correct.

All chemical reactions eventually reach equilibrium, the state at which the reactants and products are present in concentrations with no further tendency to change with time.

Therefore, the rate in the forward and reverse directions is equal.

119. A is correct.

Note the following balanced reaction and calculation:

$$2\,NH_3 \rightarrow N_2 + 3H_2$$

$$K = [\text{products}] / [\text{reactants}]$$

$$K = [N_2]\cdot[H_2]^3 / [NH_3]^2$$

$$K = (0.3)^3\cdot(0.2) / (0.1)^2$$

120. C is correct.

Usually, to solve this kind of problem, determine the order of the reactants (X, Y, Z).

However, look at the answer choices; they pertain to the order of X. Therefore, calculate the order of X.

Find 2 experiments where other reactants (Y and Z) concentrations are constant: 3 and 4.

Notice that when the concentration of X is reduced from 0.600 M to 0.200 M (i.e., a factor of 3), the rate decreases by a factor of 3 (from 15 to 5).

Therefore, the reaction is first order with respect to X.

Notes for active learning

Notes for active learning

Notes for active learning

APPENDIX

Periodic Table of the Elements

Key: atomic number / **Symbol** / name / abridged standard atomic weight

Group 1	2	3	4	5	6	7	8	9	10	11	12	13	14	15	16	17	18
1 **H** hydrogen 1.0080 ±0.0002																	2 **He** helium 4.0026 ±0.0001
3 **Li** lithium 6.94 ±0.06	4 **Be** beryllium 9.0122 ±0.0001											5 **B** boron 10.81 ±0.02	6 **C** carbon 12.011 ±0.002	7 **N** nitrogen 14.007 ±0.001	8 **O** oxygen 15.999 ±0.001	9 **F** fluorine 18.998 ±0.001	10 **Ne** neon 20.180 ±0.001
11 **Na** sodium 22.990 ±0.001	12 **Mg** magnesium 24.305 ±0.002											13 **Al** aluminium 26.982 ±0.001	14 **Si** silicon 28.085 ±0.001	15 **P** phosphorus 30.974 ±0.001	16 **S** sulfur 32.06 ±0.02	17 **Cl** chlorine 35.45 ±0.01	18 **Ar** argon 39.95 ±0.16
19 **K** potassium 39.098 ±0.001	20 **Ca** calcium 40.078 ±0.004	21 **Sc** scandium 44.956 ±0.001	22 **Ti** titanium 47.867 ±0.001	23 **V** vanadium 50.942 ±0.001	24 **Cr** chromium 51.996 ±0.001	25 **Mn** manganese 54.938 ±0.001	26 **Fe** iron 55.845 ±0.002	27 **Co** cobalt 58.933 ±0.001	28 **Ni** nickel 58.693 ±0.001	29 **Cu** copper 63.546 ±0.003	30 **Zn** zinc 65.38 ±0.02	31 **Ga** gallium 69.723 ±0.001	32 **Ge** germanium 72.630 ±0.008	33 **As** arsenic 74.922 ±0.001	34 **Se** selenium 78.971 ±0.008	35 **Br** bromine 79.904 ±0.003	36 **Kr** krypton 83.798 ±0.002
37 **Rb** rubidium 85.468 ±0.001	38 **Sr** strontium 87.62 ±0.01	39 **Y** yttrium 88.906 ±0.001	40 **Zr** zirconium 91.224 ±0.002	41 **Nb** niobium 92.906 ±0.001	42 **Mo** molybdenum 95.95 ±0.01	43 **Tc** technetium [97]	44 **Ru** ruthenium 101.07 ±0.02	45 **Rh** rhodium 102.91 ±0.01	46 **Pd** palladium 106.42 ±0.01	47 **Ag** silver 107.87 ±0.01	48 **Cd** cadmium 112.41 ±0.01	49 **In** indium 114.82 ±0.01	50 **Sn** tin 118.71 ±0.01	51 **Sb** antimony 121.76 ±0.01	52 **Te** tellurium 127.60 ±0.03	53 **I** iodine 126.90 ±0.01	54 **Xe** xenon 131.29 ±0.01
55 **Cs** caesium 132.91 ±0.01	56 **Ba** barium 137.33 ±0.01	57–71 lanthanoids	72 **Hf** hafnium 178.49 ±0.01	73 **Ta** tantalum 180.95 ±0.01	74 **W** tungsten 183.84 ±0.01	75 **Re** rhenium 186.21 ±0.01	76 **Os** osmium 190.23 ±0.03	77 **Ir** iridium 192.22 ±0.01	78 **Pt** platinum 195.08 ±0.02	79 **Au** gold 196.97 ±0.01	80 **Hg** mercury 200.59 ±0.01	81 **Tl** thallium 204.38 ±0.01	82 **Pb** lead 207.2 ±1.1	83 **Bi** bismuth 208.98 ±0.01	84 **Po** polonium [209]	85 **At** astatine [210]	86 **Rn** radon [222]
87 **Fr** francium [223]	88 **Ra** radium [226]	89–103 actinoids	104 **Rf** rutherfordium [267]	105 **Db** dubnium [268]	106 **Sg** seaborgium [269]	107 **Bh** bohrium [270]	108 **Hs** hassium [269]	109 **Mt** meitnerium [277]	110 **Ds** darmstadtium [281]	111 **Rg** roentgenium [282]	112 **Cn** copernicium [285]	113 **Nh** nihonium [286]	114 **Fl** flerovium [290]	115 **Mc** moscovium [290]	116 **Lv** livermorium [293]	117 **Ts** tennessine [294]	118 **Og** oganesson [294]

Lanthanoids

57 **La** lanthanum 138.91 ±0.01	58 **Ce** cerium 140.12 ±0.01	59 **Pr** praseodymium 140.91 ±0.01	60 **Nd** neodymium 144.24 ±0.01	61 **Pm** promethium [145]	62 **Sm** samarium 150.36 ±0.02	63 **Eu** europium 151.96 ±0.01	64 **Gd** gadolinium 157.25 ±0.03	65 **Tb** terbium 158.93 ±0.01	66 **Dy** dysprosium 162.50 ±0.01	67 **Ho** holmium 164.93 ±0.01	68 **Er** erbium 167.26 ±0.01	69 **Tm** thulium 168.93 ±0.01	70 **Yb** ytterbium 173.05 ±0.02	71 **Lu** lutetium 174.97 ±0.01

Actinoids

89 **Ac** actinium [227]	90 **Th** thorium 232.04 ±0.01	91 **Pa** protactinium 231.04 ±0.01	92 **U** uranium 238.03 ±0.01	93 **Np** neptunium [237]	94 **Pu** plutonium [244]	95 **Am** americium [243]	96 **Cm** curium [247]	97 **Bk** berkelium [247]	98 **Cf** californium [251]	99 **Es** einsteinium [252]	100 **Fm** fermium [257]	101 **Md** mendelevium [258]	102 **No** nobelium [259]	103 **Lr** lawrencium [262]

International Union of Pure and Applied Chemistry (IUPAC), 4 May 2022

Notes for active learning

Common Chemistry Equations

Throughout the test the following symbols have the definitions specified unless otherwise noted.

L, mL	= liter(s), milliliter(s)		mm Hg	= millimeters of mercury
g	= gram(s)		J, kJ	= joule(s), kilojoule(s)
nm	= nanometer(s)		V	= volt(s)
atm	= atmosphere(s)		mol	= mole(s)

ATOMIC STRUCTURE

$$E = h\nu$$

$$c = \lambda\nu$$

E = energy

ν = frequency

λ = wavelength

Planck's constant, $h = 6.626 \times 10^{-34}$ J s

Speed of light, $c = 2.998 \times 10^{8}$ m s^{-1}

Avogadro's number $= 6.022 \times 10^{23}$ mol^{-1}

Electron charge, $e = -1.602 \times 10^{-19}$ coulomb

EQUILIBRIUM

$$K_c = \frac{[C]^c[D]^d}{[A]^a[B]^b}, \text{ where } a\,A + b\,B \rightleftarrows c\,C + d\,D$$

$$K_p = \frac{(P_C)^c(P_D)^d}{(P_A)^a(P_B)^b}$$

$$K_a = \frac{[H^+][A^-]}{[HA]}$$

$$K_b = \frac{[OH^-][HB^+]}{[B]}$$

$$K_w = [H^+][OH^-] = 1.0 \times 10^{-14} \text{ at } 25°C$$

$$= K_a \times K_b$$

$$pH = -\log[H^+], \quad pOH = -\log[OH^-]$$

$$14 = pH + pOH$$

$$pH = pK_a + \log\frac{[A^-]}{[HA]}$$

$$pK_a = -\log K_a, \quad pK_b = -\log K_b$$

Equilibrium Constants

K_c (molar concentrations)

K_p (gas pressures)

K_a (weak acid)

K_b (weak base)

K_w (water)

KINETICS

$$\ln[A]_t - \ln[A]_0 = -kt$$

$$\frac{1}{[A]_t} - \frac{1}{[A]_0} = kt$$

$$t_{1/2} = \frac{0.693}{k}$$

k = rate constant

t = time

$t_{1/2}$ = half-life

GASES, LIQUIDS, AND SOLUTIONS

$$PV = nRT$$

$$P_A = P_{total} \times X_A, \text{ where } X_A = \frac{\text{moles A}}{\text{total moles}}$$

$$P_{total} = P_A + P_B + P_C + \ldots$$

$$n = \frac{m}{M}$$

$$K = {}^{\circ}C + 273$$

$$D = \frac{m}{V}$$

$$KE \text{ per molecule} = \frac{1}{2}mv^2$$

Molarity, M = moles of solute per liter of solution

$$A = abc$$

P = pressure
V = volume
T = temperature
n = number of moles
m = mass
M = molar mass
D = density
KE = kinetic energy
v = velocity
A = absorbance
a = molar absorptivity
b = path length
c = concentration

Gas constant, R = 8.314 J mol^{-1} K^{-1}

$\qquad$ = 0.08206 L atm mol^{-1} K^{-1}

$\qquad$ = 62.36 L torr mol^{-1} K^{-1}

1 atm = 760 mm Hg

$\qquad$ = 760 torr

STP = 0.00 °C and 10^5 Pa

THERMOCHEMISTRY/ ELECTROCHEMISTRY

$$q = mc\Delta T$$

$$\Delta S^{\circ} = \sum S^{\circ} \text{ products} - \sum S^{\circ} \text{ reactants}$$

$$\Delta H^{\circ} = \sum \Delta H_f^{\circ} \text{ products} - \sum \Delta H_f^{\circ} \text{ reactants}$$

$$\Delta G^{\circ} = \sum \Delta G_f^{\circ} \text{ products} - \sum \Delta G_f^{\circ} \text{ reactants}$$

$$\Delta G^{\circ} = \Delta H^{\circ} - T\Delta S^{\circ}$$

$$= -RT \ln K$$

$$= -nFE^{\circ}$$

$$I = \frac{q}{t}$$

q = heat
m = mass
c = specific heat capacity
T = temperature
S° = standard entropy
H° = standard enthalpy
G° = standard free energy
n = number of moles
E° = standard reduction potential
I = current (amperes)
q = charge (coulombs)
t = time (seconds)

Faraday's constant, F = 96,485 coulombs per mole of electrons

$$1 \text{ volt} = \frac{1 \text{ joule}}{1 \text{ coulomb}}$$

154

Annotated Glossary of Chemistry Terms

A

Absolute entropy (of a substance) – the increase in the entropy of a substance as it goes from a perfectly ordered crystalline form at 0 K (where its entropy is zero) to the temperature in question.

Absolute zero – the zero point on the absolute temperature scale; –273.15 °C or 0 K; theoretically, the temperature at which molecular motion ceases (i.e., the system does not emit or absorb energy, atoms at rest).

Absorption spectrum – spectrum associated with the absorption of electromagnetic radiation by atoms (or other species), resulting from transitions from lower to higher energy states.

Accuracy – the degree to which a value is close to the actual value; see *precision*.

Acid – a substance that produces H^+ (*aq*) ions in an aqueous solution and gives a pH of less than 7.0; strong acids ionize entirely or almost entirely in dilute aqueous solution; weak acids ionize only slightly. It turns litmus red.

Acid dissociation constant – an equilibrium constant for dissociating a weak acid.

Acid rain – rainwater with a pH of less than 5.7; caused by the gases NO_2 (vehicle exhaust fumes) and SO_2 (from burning fossil fuels) dissolving in the rain. It kills fish, wildlife, and trees and destroys buildings and lakes.

Acidic salt – contains an ionizable hydrogen atom; does not necessarily produce acidic solutions.

Actinides – the fifteen chemical elements that are between actinium (89) and lawrencium (103).

Activated complex – a structure formed by a collision between molecules in which new bonds form.

Activation energy – the amount of energy that reactants must absorb in their ground states to reach the transition state needed for a reaction to occur.

Active metal – a metal with low ionization energy that loses electrons readily to form cations.

Activity (of a component of ideal mixture) – a dimensionless quantity whose magnitude is equal to the molar concentration in an ideal solution; equal to partial pressure in an ideal gas mixture; 1 for pure solids or liquids.

Activity series – a listing of metals (and hydrogen) in order of decreasing activity.

Actual yield – the amount of a specified pure product obtained from a given reaction; see *theoretical yield*.

Addition reaction – a reaction in which two atoms or groups of atoms are added to a molecule, one on each side of a double or triple bond.

Adhesive forces – forces of attraction between a liquid and another surface.

Adsorption – the adhesion of a species onto the surfaces of particles.

Aeration – the mixing of air into a liquid or a solid.

Alcohol – a hydrocarbon derivative containing a ~OH group attached to a carbon atom, not in an aromatic ring.

Alkali metals – the elements of Group IA on the periodic table (e.g., Na, K, Rb).

Alkaline battery – a dry cell in which the electrolyte contains KOH.

Alkaline earth metals – group IIA metals on the periodic table; see *earth metals*.

Allomer – a substance that has a different composition from another but the same crystalline structure.

Allotropes – elements with different structures (therefore different forms), such as carbon (e.g., diamond, graphite, and fullerene).

Allotropic modifications (allotropes) – different forms of the same element in the same physical state.

Alloy – a mixture of metals. For example, bronze is an alloy formed from copper and tin.

Alloying – mixing metals with other substances (usually other metals) to modify their properties.

Alpha (α) particle – a helium nucleus; helium ion with 2+ charge; an assembly of two protons and two neutrons.

Amorphous solid – a non-crystalline solid with no well-defined ordered structure.

Ampere – unit of electrical current; one ampere equals one coulomb per second.

Amphiprotism – the ability of a substance to exhibit amphiprotic by accepting donated protons.

Amphoterism – the ability to react with both acids and bases; to act as either an acid or a base.

Amplitude – the maximum distance that medium particles carrying the wave move from their rest position.

Anion – a negative ion; an atom or group of atoms that has gained one or more electrons.

Anode – in a cathode ray tube, the positive electrode (electrode at which oxidation occurs); the positive side of a dry cell battery or a cell.

Antibonding orbital – a molecular orbital higher in energy than any of the atomic orbitals from which it is derived; lends instability to a molecule or ion when populated with electrons; denoted with star (*) superscript.

Artificial transmutation – an artificially induced nuclear reaction caused by the bombardment of a nucleus with subatomic particles or small nuclei.

Associated ions – short-lived species formed by the collision of dissolved ions of opposite charges.

Atmosphere – a unit of pressure; the pressure supports a column of mercury 760 mm high at 0 °C.

Atom – a chemical element in its smallest form; it comprises neutrons and protons within the nucleus and electrons circling the nucleus; it is the smallest part of an element that can exist.

Atomic mass unit (amu) – one-twelfth of the mass of an atom of the carbon-12 isotope; used for stating atomic and formula weights; known as a dalton.

Atomic number – represents an element corresponding with the number of protons within the nucleus; the number of protons in the nucleus of the atom.

Atomic orbital (*AO*) – a region or volume in space where the probability of finding electrons is highest.

Atomic radius – radius of an atom.

Atomic weight – weighted average of the masses of the constituent isotopes of an element; the relative masses of atoms of different elements.

Aufbau (or *building up*) principle – describes the order in which electrons fill orbitals in atoms.

Autoionization – an ionization reaction between identical molecules.

Avogadro's Law – equal volumes of gases contain the same number of molecules at the same temperature and pressure.

Avogadro's number (*N*$_A$) – the number (6.022×10^{23}) of atoms, molecules, or particles found in precisely 1 mole of a substance.

B

Background radiation – extraneous to an experiment; usually the low-level natural radiation from cosmic rays and trace radioactive substances present in the environment.

Band – a series of very closely spaced nearly continuous molecular orbitals that belong to the crystal as a whole.

Band of stability – band containing nonradioactive nuclides in a plot of neutrons *vs.* their atomic number.

Band theory of metals – the theory that accounts for the bonding and properties of metallic solids.

Barometer – a device used to measure the pressure in the atmosphere.

Base – a substance that produces $^-$OH (*aq*) ions in an aqueous solution; accepts a proton and has a high pH; strongly soluble bases are soluble in water and are entirely dissociated; weak bases ionize only slightly; a typical example of a base is sodium hydroxide (NaOH). It turns litmus blue.

Basic anhydride – the oxide of a metal that reacts with water to form a base.

Basic salt – a salt containing an ionizable OH group.

Beta (β) particle – an electron emitted from the nucleus when a neutron decays to a proton and an electron.

Binary acid – a binary compound in which H is bonded to one or more electronegative nonmetals.

Binary compound – consists of two elements; it may be ionic or covalent.

Binding energy (nuclear binding energy) – the energy equivalent ($E = mc^2$) of the mass deficiency of an atom (where E is the energy in joules, m is the mass in kilograms, and c is the speed of light in m/s^2).

Boiling – the phase transition of a liquid vaporizing.

Boiling point – the temperature at which the vapor pressure of a liquid is equal to the applied pressure; the *condensation point*.

Boiling point elevation – the increase in the boiling point of a solvent caused by the dissolution of a nonvolatile solute.

Bomb calorimeter – a device used to measure the heat transfer between a system and its surroundings at constant volume.

Bond – the attraction and repulsion between atoms and molecules is a cornerstone of chemistry.

Bond energy – the amount of energy necessary to break one mole of bonds in a substance, dissociating the substance in its gaseous state into atoms of its elements in the gaseous state.

Bond order – half the number of electrons in bonding orbitals minus half the electrons in antibonding orbitals.

Bonding orbital – a molecular orbit lower in energy than any of the atomic orbitals from which it is derived; lends stability to a molecule or ion when populated with electrons.

Bonding pair – pair of electrons involved in a covalent bond.

Boron hydrides – binary compounds of boron and hydrogen.

Born-Haber cycle – a series of reactions (and the accompanying enthalpy changes) which, when summed, represent the hypothetical one-step reaction by which elements in their standard states are converted into crystals of ionic compounds (and the accompanying enthalpy changes).

Boyle's Law – at a constant temperature, the volume occupied by a definite mass of a gas is inversely proportional to the applied pressure.

Breeder reactor – a nuclear reactor that produces more fissionable nuclear fuel than it consumes.

Brønsted-Lowrey acid – a chemical species that donates a proton.

Brønsted-Lowrey base – a chemical species that accepts a proton.

Buffer solution – resists change in pH; contains either a weak acid and a soluble ionic salt of the acid or a weak base and a soluble ionic salt of the base.

Buret – a piece of volumetric glassware, usually graduated in 0.1 mL intervals, used to deliver solutions for titrations in a quantitative (drop-like) manner; also spelled *burette*.

C

Calorie – the amount of heat required to raise the temperature of one gram of water from 14.5 °C to 15.5 °C; 1 calorie = 4.184 joules.

Calorimeter – a device used to measure the heat transfer between a system and its surroundings.

Canal ray – a stream of positively charged particles (cations) that moves toward the negative electrode in cathode ray tubes; observed to pass through canals in the negative electrode.

Capillary – a tube having a very small inside diameter.

Capillary action – the drawing of a liquid up the inside of a small-bore tube when adhesive forces exceed cohesive forces; the depression of the surface of the liquid when cohesive forces exceed the adhesive forces.

Catalyst – a chemical compound used to change the rate (to speed or slow it) of a regenerated reaction (i.e., not consumed) at the end of the reaction.

Catenation – the bonding of atoms of the same element into chains or rings (i.e., the ability of an element to bond with itself).

Cathode – the electrode at which reduction occurs; in a cathode ray tube, the negative electrode.

Cathodic protection – protection of a metal (making a cathode) against corrosion by attaching it to a sacrificial anode of a more easily oxidized metal.

Cathode ray tube – a closed glass tube containing gas under low pressure, with electrodes near the ends and a luminescent screen near the positive electrode; produces cathode rays when a high voltage is applied.

Cation – a positive ion; an atom or group of atoms that has lost one or more electrons.

Cell potential – the potential difference, E_{cell}, between oxidation and reduction half-cells under nonstandard conditions; the force in a galvanic cell pulls electrons through a reducing agent to an oxidizing agent.

Central atom – an atom in a molecule or polyatomic ion bonded to more than one other atom.

Chain reaction – a reaction that, once initiated, sustains itself and expands; a reaction in which reactive species, such as radicals, are produced in more than one step, allowing these reactive species to propagate the chain reaction.

Charles' Law – at constant pressure, the volume occupied by a definite mass of gas is directly proportional to its absolute temperature.

Chemical bonds – the attractive forces holding atoms together in elements or compounds.

Chemical change – when one or more new substances are formed.

Chemical equation – description of a chemical reaction by placing the formulas of the reactants on the left of an arrow and the formulas of the products on the right.

Chemical equilibrium – a state of dynamic balance in which the rates of forward and reverse reactions are equal; there is no net change in concentrations of reactants or products while a system is at equilibrium.

Chemical kinetics – studies rates and mechanisms of chemical reactions and factors they depend on.

Chemical periodicity – the variations in properties of elements with their position in the periodic table.

Chemical reaction – the change of one or more substances into another or multiple substances.

Cloud chamber – a device for observing the paths of speeding particles as vapor molecules condense on them to form fog-like tracks.

Cobalt chloride paper – water test; water changes the color from blue to pink.

Coefficient of expansion – the ratio of the change in the length or the volume of a body to the original length or volume for a unit change in temperature.

Cohesive forces – the forces of attraction among particles of a liquid.

Colligative properties – physical properties of solutions that depend upon the number but not the kind of solute particles present.

Collision theory – the theory of reaction rates that states that effective collisions between reactant molecules must occur for the reaction to occur.

Colloid – a heterogeneous mixture in which solute-like particles do not settle (e.g., many kinds of milk).

Combination reaction – two substances (elements or compounds) combine to form one compound.

Combustible – classification of liquid substances that burn based on flashpoints; any liquid having a flashpoint at or above 37.8 °C (100 °F) but below 93.3 °C (200 °F), except any mixture having components with flashpoints of 93.3 °C (200 °F) or higher, the total of which makes up 99% or more of the volume of the mixture.

Combustion (or *burning*) – an exothermic reaction between an oxidant and fuel with heat and often light.

Common ion effect – suppression of ionization of a weak electrolyte by the presence in the same solution of a strong electrolyte containing one of the same ions as the weak electrolyte.

Complex ions – ions resulting from coordinating covalent bonds between simple ions and other ions or molecules.

Composition stoichiometry – describes the quantitative (mass) relationships among elements in compounds.

Compound – a substance of two or more chemically bonded elements in fixed proportions; can be decomposed into constituent elements.

Compressed gas – a single or mixture of gases having (in a container) an absolute pressure exceeding 40 psi at 21.1 °C (70 °F).

Compression – an area in a longitudinal wave where the particles are closer and pushed in.

Concentration – the amount of solute per unit volume, the mass of solvent or solution.

Condensation – the phase change from gas to liquid.

Condensed phases – the liquid and solid phases; phases in which particles interact strongly.

Condensed states – the solid and liquid states.

Conduction – heat transfer between substances in direct contact with each other (i.e., must be touching); when particles of a hotter substance vibrate, these molecules bump into nearby particles and transfer some energy.

Conduction band – a vacant or partially filled band of energy levels just higher in energy than a filled band; a band within which, or into which, electrons must be promoted to allow electrical conduction to occur in a solid.

Conductor – a material that allows electric flow more freely.

Conjugate acid-base pair – in Brønsted-Lowry terms, reactant and product that differ by a proton, H^+.

Conformations – structures of a compound that differ by the extent of rotation about a single bond.

Continuous spectrum – contains wavelengths in a specified region of the electromagnetic spectrum.

Control rods – rods of materials such as cadmium or boron steel that act as neutron absorbers (not merely moderators), used in nuclear reactors to control neutron fluxes and therefore fission rates.

Conjugated double bonds – double bonds separated from each other by one single bond –C=C–C=C–

Contact process – the industrial process for sulfur trioxide (SO_3) and sulfuric acid (H_2SO_4) production from sulfur dioxide (SO_2).

Convection – the physical flow of matter when heat flows by energized molecules from one place to another through the movement of fluids. Transfer of heat through a liquid or a gas occurs when molecules of the liquid or gas move and carry the heat.

Coordinate covalent bond – a covalent bond with shared electrons furnished by the same species; a bond between a Lewis acid and a Lewis base.

Coordination compound or complex – a compound containing coordinate covalent bonds.

Coordination number – the number of donor atoms coordinated to a metal; the number of nearest neighbors of an atom or ion in describing crystals.

Coordination sphere – the metal ion and its coordinating ligands, but no uncoordinated counter-ions.

Corrosion – oxidation of metals (e.g., rusting) in the presence of air and moisture.

Coulomb – the SI unit of electrical charge; unit symbol – C.

Covalent bond – a force of attraction (chemical bond) formed by the sharing of electron pairs between two atoms.

Covalent compounds – compounds made of two or more nonmetal atoms bonded by sharing valence electrons.

Critical mass – the minimum mass of a particular fissionable nuclide in a given volume required to sustain a nuclear chain reaction.

Critical point – the combination of critical temperature and critical pressure of a substance.

Critical pressure – the pressure required to liquefy a gas (vapor) at its *Critical temperature*.

Critical temperature – the temperature above which a gas cannot be liquefied; the temperature above which a substance cannot exhibit distinct gas and liquid phases.

Crystal – a solid packed with ions, molecules, or atoms in an orderly fashion.

Crystal field stabilization energy – a measure of the net energy of stabilization gained by a metal ion's nonbonding d electrons due to complex formation.

Crystal field theory – bonding in transition metal complexes in which ligands and metal ions are treated as point charges; a purely ionic model; ligand point charges represent the crystal (electrical) field perturbing the metal's d orbitals containing nonbonding electrons.

Crystal lattice – a pattern of arrangement of particles in a crystal.

Crystal lattice energy – the amount of energy that holds a crystal together; the energy change when a mole of solid forms from its constituent molecules or ions (for ionic compounds) in their gaseous state (consistently negative).

Crystalline solid – a solid characterized by a regular, ordered arrangement of particles.

Curie (Ci) – the basic unit to describe the intensity of radioactivity in a sample of material; one curie equals 37 billion disintegrations per second or approximately the amount of radioactivity given off by 1 gram of radium.

Current – a flow of charged particles, such as electrons or ions, moving through an electrical space or conductor. It is measured as the net rate of flow of electric charge; the unit is an Ampere (A).

Cuvette – glassware used in spectroscopic experiments; usually made of plastic, glass, or quartz, and should be as clean and transparent as possible.

Cyclotron – a device for accelerating charged particles along a spiral path.

D

Dalton's Law (or the *law of partial pressures*) – the pressure exerted by a mixture of gases is the sum of the partial pressures of the individual gases.

Daughter nuclide – nuclide produced in nuclear decay.

Debye – the unit used to express dipole moments.

Degenerate – in orbitals, describes orbitals of the same energy.

Deionization – the removal of ions; in the case of water, mineral ions such as sodium, iron, and calcium.

Deliquescence – substances that absorb water from the atmosphere to form liquid solutions.

Delocalization – in reference to electrons, bonding electrons distributed among more than two atoms bonded; occurs in species that exhibit resonance.

Density – mass per unit volume; $D = M \times V$.

Deposition – settling particles within a solution; the direct solidification of vapor by cooling; see *sublimation*.

Derivative – a compound that can be imagined arising from a parent compound by replacing one atom with another atom or group of atoms; used extensively in organic chemistry to identify compounds.

Detergent – a soap-like emulsifier with sulfate, SO_3, or a phosphate group instead of carboxylate group.

Deuterium – an isotope of hydrogen whose atoms are twice as massive as ordinary hydrogen; deuterium atoms contain a proton and a neutron in the nucleus.

Dextrorotatory – refers to an optically active substance that rotates plane-polarized light clockwise, also known as *dextro*.

Diagonal similarities – chemical similarities in the Periodic Table of Elements of Period 2 to elements of Period 3 one group to the right, especially evident toward the left of the periodic table.

Diamagnetism – weak repulsion by a magnetic field.

Differential Scanning Calorimetry (DSC) – a technique for measuring temperature, direction, and magnitude of thermal transitions in a sample material by heating/cooling and comparing the amount of energy required to maintain its rate of temperature increase or decrease with an inert reference material under similar conditions.

Differential Thermal Analysis (DTA) – a technique for observing the temperature, direction, and magnitude of thermally induced transitions in a material by heating/cooling a sample and comparing its temperature with an inert reference material under similar conditions.

Differential thermometer – a thermometer used to measure minimal temperature changes accurately.

Dilution – the process of reducing the concentration of a solute in a solution, usually by mixing with more solvent.

Dimer – a molecule formed by combining two smaller (identical) molecules.

Dipole – electric or magnetic separation of charge; charge separation between two covalently bonded atoms.

Dipole-dipole interactions – attractive electrostatic forces between polar molecules (i.e., between molecules with permanent dipoles).

Dipole moment – the product of the distance separating opposite charges of equal magnitude; a measure of the polarity of a bond or molecule; a measured dipole refers to the dipole moment of an entire molecule.

Dispersing medium – the solvent-like phase in a colloid.

Dispersed phase – the solute-like species in a colloid.

Displacement reactions – reactions in which one element displaces another from a compound.

Disproportionation reactions – redox reactions in which the oxidizing agent and the reducing agent are the same species.

Dissociation – in an aqueous solution, the process by which a solid ionic compound separates into its ions.

Dissociation constant – equilibrium constant for dissociating a complex ion into a simple ion and coordinating species (ligands).

Dissolution or solvation – the spread of ions in a monosaccharide.

Distilland – the material in a distillation apparatus that is to be distilled.

Distillate – the material in a distillation apparatus collected in the receiver.

Distillation – separating a liquid mixture into its components based on differences in boiling points; the process in which components of a mixture are separated by boiling away the more volatile liquid; the vaporization of a liquid by heating and then the condensation of the vapor by cooling.

Domain – a cluster of atoms in a ferromagnetic substance, which align in the same direction in the presence of an external magnetic field.

Donor atom – a ligand atom whose electrons are shared with a Lewis acid.

d-orbitals – beginning in the third energy level, a set of five degenerate orbitals per energy level, higher in energy than *s* and *p* orbitals of the same energy level.

Dosimeter – a small, calibrated electroscope worn by laboratory personnel to measure incident ionizing radiation or chemical exposure.

Double bond – covalent bond resulting from the sharing of four electrons (two pairs) between two atoms.

Double salt – a solid consisting of two co-crystallized salts.

Doublet – two peaks or bands of about equal intensity appearing close on a spectrogram.

Downs cell – an electrolytic cell for the commercial electrolysis of molten sodium chloride.

DP number – the degree of polymerization; the average number of monomer units per polymer unit.

Dry cells (voltaic cells) – ordinary batteries for appliances (e.g., flashlights, radios).

Dumas method – a method used to determine the molecular weights of volatile liquids.

Dynamic equilibrium – an equilibrium in which the processes occur continuously with no net change.

E

Earth metal – highly reactive elements in group IIA of the periodic table (includes beryllium, magnesium, calcium, strontium, barium, and radium); see *alkaline earth metal*.

Effective collisions – a collision between molecules resulting in a reaction, one in which the molecules collide with proper relative orientations and sufficient energy to react.

Effective molality – the sum of the molalities of solute particles in a solution.

Effective nuclear charge – the nuclear charge experienced by the outermost electrons of an atom; the actual nuclear charge minus the effects of shielding due to inner-shell electrons (e.g., a set of $d_{x^2-y^2}$ and d_{z^2} orbitals); those d orbitals within a set with lobes directed along the x, y, and z axes.

Electrical conductivity – the measure of how easily an electric current can flow through a substance.

Electric charge – a measured property (coulombs) that determines electromagnetic interaction.

Electrochemical cell – using a chemical reaction, an electromotive force is generated.

Electrochemistry – the study of chemical changes produced by electrical current and electricity production by chemical reactions.

Electrodes – surfaces upon which oxidation and reduction half-reactions occur in electrochemical cells; a conductor dips into an electrolyte and allows the electrons to flow to and from the electrolyte.

Electrode potentials – potentials, E, of half-reactions as reductions vs. standard hydrogen electrode.

Electrolysis – 1) occurs in electrolytic cells; chemical decomposition occurs by passing an electric current through a solution containing ions. 2) producing a chemical change using electricity; used to split up water into H and O_2.

Electrolyte – a substance (i.e., anions, cations) which, when dissolved in water, conducts electricity. An ionic solution that conducts a certain amount of current is categorized into two types: weak and strong.

Electrolytic cells – electrochemical cells in which electrical energy causes nonspontaneous redox reactions to occur (i.e., forced to occur by applying an outside source of electrical energy).

Electrolytic conduction – electrical current passes through ions in a solution or pure liquid.

Electromagnetic radiation – energy propagated using electric and magnetic fields that oscillate in directions perpendicular to the direction of travel of the energy; a type of wave that can go through vacuums as well as material; classified as a "self-propagating wave."

Electromagnetism – fields of an electric charge and electric properties that change how particles move and interact.

Electromotive force – a device that gains energy as electric charges pass through it.

Electromotive series – the relative order of tendencies for elements and their simple ions to act as oxidizing or reducing agents; also known as the "activity series."

Electron – a subatomic particle having a mass of 0.00054858 amu and a charge of –1.

Electron affinity – the energy absorbed in the process in which an electron is added to a neutral isolated gaseous atom to form a gaseous ion with a 1– charge; it has a negative value if energy is released.

Electron configuration – the specific distribution of electrons in atomic orbitals of atoms or ions.

Electron-deficient compounds – at least one atom (other than H) shares fewer than eight electrons.

Electron shells – an orbital around an atom's nucleus with a fixed number of electrons (usually 2 or 8).

Electronic transition – the transfer of an electron from one energy level to another.

Electronegativity – a measure of the relative tendency of an atom to attract electrons to itself when chemically combined with another atom.

Electronic geometry – the geometric arrangement of orbitals containing the shared and unshared electron pairs surrounding the central atom or polyatomic ion.

Electrophile – positively charged or electron-deficient.

Electrophoresis – separating ions by migration rate and direction of migration in an electric field.

Electroplating – a metal is covered with another metal layer using electricity; plating a metal onto a (cathodic) surface by electrolysis.

Element – a substance that cannot be decomposed into simpler substances by chemical means; defined by its *atomic number*. A substance that cannot be split into simpler substances by chemical means.

Eluant (or eluent) – the solvent used in the process of elution, as in liquid chromatography.

Eluate – a solvent (or mobile phase) which passes through a chromatographic column and removes the sample components from the stationary phase.

Emission spectrum – the emission of electromagnetic radiation by atoms (or other species) resulting from electronic transitions from higher to lower energy states.

Empirical formula – the simplest whole-number ratio of atoms of each element present in a compound; also known as the *simplest formula*.

Emulsifying agent – a substance that coats the particles of the dispersed phase and prevents coagulation of colloidal particles; an emulsifier.

Emulsion – colloidal suspension of a liquid in a liquid.

Endothermic – describes processes that absorb heat energy (H).

Endothermicity – the absorption of heat by a system as the process occurs.

Endpoint – the point at which an indicator changes color and titration stops.

Energy – a system's ability to do work.

Enthalpy (*H*) – the heat content of a specific amount of substance; $E = PV$.

Entropy *(S)* – a thermodynamic state or property that measures the degree of disorder (i.e., randomness) of a system; the amount of energy not available for work in a closed thermodynamic system (usually denoted by S).

Enzyme – a protein that acts as a catalyst in biological systems.

Equation of state – an expression that describes the behavior of matter in a given state; the van der Waals equation describes the behavior of the gaseous state.

Equilibrium or chemical equilibrium – a state of dynamic balance with the rates of forward and reverse reactions equal; the state of a system where neither forward nor reverse reaction is thermodynamically favored.

Equilibrium constant – a quantity characterizing equilibrium position for a reversible reaction; its magnitude is equal to mass action expression at equilibrium; equilibrium, "K," varies with temperature.

Equivalence point – the point when chemically equivalent amounts of reactants have reacted.

Equivalent weight – an oxidizing or reducing agent whose mass gains (oxidizing agents) or loses (reducing agents) 6.022×10^{23} electrons in a redox reaction.

Evaporation – vaporization of a liquid below its boiling point.

Evaporation rate – the rate at which a particular substance vaporizes (evaporate) compared to the rate of a known substance, such as ethyl ether, especially useful for health and fire-hazard considerations.

Excited state – any state other than the ground state of an atom or molecule; see *ground state*.

Exothermic – describes processes that release heat energy (*H*).

Exothermicity – the release of heat by a system as a process occurs.

Explosive – a chemical or compound that causes a sudden, almost instantaneous release of pressure, gas, heat, and light when subjected to sudden shock, pressure, high temperature, or applied potential.

Explosive limits – the range of concentrations over which a flammable vapor mixed with the proper ratios of air will ignite or explode if a source of ignition is provided.

Extensive property – a property that depends upon the amount of material in a sample.

Extrapolate – to estimate the value of a result outside the range of a series of known values; a technique used in the standard additions calibration procedure.

F

Faraday constant – a unit of electrical charge widely used in electrochemistry and equal to ~ 96,500 coulombs; represents 1 mole of electrons, or the Avogadro number of electrons: 6.022×10^{23} electrons.

Faraday's law of electrolysis – a two-part law that Michael Faraday published about electrolysis. 1. the mass of a substance altered at an electrode during electrolysis is directly proportional to the quantity of electricity transferred at that electrode. 2. the mass of an elemental material altered at an electrode is directly proportional to the element's equivalent weight; one equivalent weight of a substance is produced at each electrode during the passage of 96,487 coulombs of charge through an electrolytic cell.

Fast neutron – a neutron ejected at high kinetic energy in a nuclear reaction.

Ferromagnetism – the ability of a substance to become permanently magnetized by exposure to an external magnetic field.

Flashpoint – the temperature at which a liquid yields enough flammable vapor to ignite; various recognized industrial testing methods exist; therefore, the method used must be specified.

Fluorescence – absorption of high-energy radiation and subsequent emission of visible light.

First Law of Thermodynamics – the amount of energy in the universe is constant (i.e., energy is neither created nor destroyed in ordinary chemical reactions and physical changes); known as the Law of Conservation of Energy.

Fluids – substances that flow freely; gases and liquids.

Flux – a substance added to react with the charge or a product of its reduction; in metallurgy, it is usually added to lower the melting point.

Foam – colloidal suspension of a gas in a liquid.

Formal charge – a method of counting electrons in a covalently bonded molecule or ion; it counts bonding electrons as though they were equally shared between the two atoms.

Formula – a combination of symbols that indicates the chemical composition of a substance.

Formula unit – the smallest repeating unit of a substance; the molecule for nonionic substances.

Formula weight – the mass of one formula unit of a substance in atomic mass units.

Fossil fuels – formed from the remains of plants and animals that lived millions of years ago.

Fractional distillation – when a fractioning column is used in a distillation apparatus to separate the components of a liquid mixture with different boiling points.

Fractional precipitation – removal of some ions from a solution by precipitation while leaving other ions with similar properties in the solution.

Free energy change – the indicator of the spontaneity of a process at constant T and P (e.g., if ΔG is negative, the process is spontaneous).

Free radical – a highly reactive chemical species carrying no charge and having a single unpaired electron in an orbital.

Freezing – phase transition from liquid to solid.

Freezing point depression – the decrease in the freezing point of a solvent caused by the presence of a solute.

Frequency – the number of repeating points on a wave that passes a given observation point per unit time; the unit is 1 hertz = 1 cycle per 1 second.

Fuel – any substance that burns in oxygen to produce heat.

Fuel cells – a voltaic cell that converts the chemical energy of a fuel and an oxidizing agent directly into electrical energy continuously.

G

Gamma (γ) ray – a highly penetrating type of nuclear radiation similar to X-ray radiation, except that it comes from within the nucleus of an atom and has higher energy; energy-wise, very similar to cosmic rays, except that cosmic rays originate from outer space.

Galvanic cell – a battery made from an electrochemical cell with two metals connected by a salt bridge.

Galvanizing – placing a thin layer of zinc on a ferrous material to protect the underlying surface from corrosion.

Gangue – sand, rock, and other impurities surrounding the mineral of interest in an ore.

Gas – a state of matter in which the particles have no definite shape or volume, though they fill their container.

Gay-Lussac's Law – the expression used for each of the two relationships named after the French chemist Joseph Louis Gay-Lussac concerning the properties of gases; more usually applied to his law of combining volumes.

Geiger counter – a gas-filled tube that discharges electrically when ionizing radiation passes through it.

Gel – colloidal suspension of a solid dispersed in a liquid; a semi-rigid solid.

Gibbs (free) energy – the thermodynamic state function of a system that indicates the amount of energy available for the system to do useful work at constant T and P; a value that indicates the spontaneity of a reaction (usually denoted by G).

Graham's Law – the rates of effusion of gases are inversely proportional to the square roots of their molecular weights or densities.

Ground state – the lowest energy state or most stable state of an atom, molecule, or ion; see *excited state*.

Group – a vertical column in the periodic table; known as a family.

H

Haber process – a process for the catalyzed industrial production of ammonia from N_2 and H_2 at high temperature and pressure.

Half-cell – the compartment in which the oxidation or reduction half-reaction occurs in a voltaic cell.

Half-life – the time required for half of a reactant to be converted into product(s); the time required for half of a given sample to undergo radioactive decay.

Half-reaction – the oxidation or the reduction part of a redox reaction.

Halogens – group VIIA elements: F, Cl, Br, I; halogens are nonmetals.

Hard water – water high in dissolved minerals that makes it difficult to form lather with soap.

Heat – a form of energy that flows between two samples of matter because of temperature differences.

Heat capacity – the amount of heat required to raise the temperature of a mass one degree Celsius.

Heat of condensation – the amount of heat that must be removed from one gram of vapor at its condensation point to condense the vapor with no change in temperature.

Heat of crystallization – the amount of heat that must be removed from one gram of a liquid at its freezing point to freeze it with no change in temperature.

Heat of fusion – the amount of heat required to melt one gram of a solid at its melting point with no change in temperature; usually expressed in J/g; the molar heat of fusion is the amount of heat required to melt one mole of a solid at its melting point with no change in temperature and is usually expressed in kJ/mol.

Heat of solution – the amount of heat absorbed in forming a solution that contains one mole of the solute; the value is positive if heat is absorbed (endothermic) and negative if heat is released (exothermic).

Heat of vaporization – the amount of heat required to vaporize one gram of a liquid at its boiling point with no change in temperature; usually expressed in J/g; the molar heat of vaporization is the amount of heat required to vaporize one mole of liquid at its boiling point with no change in temperature and is usually expressed as ion kJ/mol.

Heisenberg uncertainty principle – states that it is impossible to determine the momentum and position of an electron simultaneously with absolute accuracy.

Henry's Law – the gas pressure above a solution is proportional to concentration of the gas in solution.

Hess' Law of heat summation – the enthalpy change for a reaction is the same whether it occurs in one step or a series of steps.

Heterogeneous catalyst – exists in a different phase (solid, liquid, or gas) from the reactants; a contact catalyst.

Heterogeneous equilibria – equilibria involving species in more than one phase.

Heterogeneous mixture – a mixture that does not have uniform composition and properties throughout.

Heteronuclear – consisting of different elements.

High spin complex – crystal field designation for an outer orbital complex; t_{2g} and e_g orbitals are singly occupied before pairing occurs.

Homogeneous catalyst – in the same phase (solid, liquid, or gas) as the reactants.

Homogeneous equilibria – when *reagents* and *products* are in the same phase (gases, liquids, or solids).

Homogeneous mixture – a mixture which has uniform composition and properties throughout.

Homologous series – compounds with each member differing from the next by a specific number and kind of atoms.

Homonuclear – consisting of only one element.

Hund's rule – single electrons must occupy orbitals of a given sublevel before pairing begins; see *Aufbau* (or *building up*) *principle*.

Hybridization – mixing atomic orbitals to form a new set of atomic orbitals with the same electron capacity and properties and energies intermediate between the original unhybridized orbitals.

Hydrate – a solid compound that contains a definite percentage of bound water.

Hydrate isomers – crystalline complexes that differ in whether water exists inside or outside the coordination sphere.

Hydration – the reaction of a substance with water.

Hydration energy – the energy change accompanying the hydration of a mole of gas and ions.

Hydride – a binary compound of hydrogen.

Hydrocarbons – compounds that contain only carbon and hydrogen.

Hydrogen bond – a relatively strong dipole-dipole interaction (but still considerably weaker than the covalent or ionic bonds) between molecules containing hydrogen directly bonded to a small, highly electronegative atom, such as N, O or F.

Hydrogenation – the reaction in which hydrogen adds across a double or triple bond.

Hydrogen-oxygen fuel cell – hydrogen is the fuel (reducing agent), and oxygen is the oxidizing agent.

Hydrolysis – the reaction of a substance with water or its ions.

Hydrolysis constant – an equilibrium constant for a hydrolysis reaction.

Hydrometer – a device used to measure the densities of liquids and solutions.

Hydrophilic colloids – colloidal particles that repel water molecules.

I

Ideal gas – a hypothetical gas that obeys the postulates of the Kinetic Molecular Theory.

Ideal gas law – the product of pressure and the volume of an ideal gas is directly proportional to the number of moles of the gas and the absolute temperature. $PV = nRT$

Ideal solution – obeys Raoult's Law strictly.

Immiscible liquids – do not mix to form a solution (e.g., oil and water).

Indicators – for acid-base titrations, organic compounds that exhibit different colors in solutions of different acidities, determine the point at which the reaction between two solutes is complete.

Inert pair effect – characteristic of the post-transition minerals; the tendency of the electrons in the outermost atomic *s* orbital to remain un-ionized or unshared in compounds of post-transition metals.

Inhibitory catalyst – an inhibitor; a catalyst that decreases the rate of reaction.

Inner orbital complex – valence bond designation for a complex in which the metal ion utilizes d orbitals for one shell inside the outermost occupied shell in its hybridization.

Inorganic chemistry – a part of chemistry concerned with inorganic (non-carbon-based) compounds.

Insulator – a material that resists the flow of electric current or heat transfer; it does not allow heat to flow easily.

Insoluble compound – a substance that will not dissolve in a solvent, even after mixing.

Integrated rate equation – an expression giving the concentration of a reactant remaining after a specified time; it has a different mathematical form for different orders of reactants.

Intermolecular forces – forces between individual particles (atoms, molecules, ions) of a substance.

Ion – a molecule that has gained or lost electrons; an atom or a group of atoms carries an electric charge (Na^+).

Ion product for water – equilibrium constant for water ionization; $Kw = [H_3O^+] \cdot [^-OH] = 1.00 \times 10^{-14}$ at 25 °C.

Ionic bond – the electrostatic attraction between oppositely charged ions, resulting from a transfer of electrons.

Ionic bonding – chemical bonding resulting from transferring electrons from one atom or group.

Ionic compounds – compounds containing predominantly ionic bonding.

Ionic geometry – arrangement of atoms (not lone pairs of electrons) about the central atom of a polyatomic ion.

Ionization – the breaking up of a compound into separate ions; in an aqueous solution, the process by which a molecular compound reacts with water and forms ions.

Ionization constant – equilibrium constant for the ionization of a weak electrolyte.

Ionization energy – the minimum amount of energy required to remove the most loosely held electron of an isolated gaseous atom or ion.

Ion exchange – a method of removing hardness from water, it replaces the positive ions that cause the hardness with H^+ ions.

Ionization isomers – result from the interchange of ions inside and outside the coordination sphere.

Isoelectric – having the same electronic configurations.

Isomers – different substances with the same formula.

Isomorphous – refers to crystals having the same atomic arrangement.

Isotopes – two or more forms of atoms of the same element with different masses; atoms containing the same number of protons but different numbers of neutrons.

IUPAC – acronym for "International Union of Pure and Applied Chemistry."

J

Joule (J) – a unit of energy in the SI system; one joule is $1 \text{ kg} \cdot \text{m}^2/\text{s}^2$, which is 0.2390 calories.

K

K capture – absorption of a K shell ($n = 1$) electron by a proton as it is converted to a neutron.

Kelvin – a unit of measure for temperature based upon an absolute scale.

Kinetics – a subfield of chemistry specializing in reaction rates.

Kinetic energy (*KE*) – energy that matter processes through its motion.

Kinetic Molecular Theory – a theory that attempts to explain macroscopic observations on gases in microscopic or molecular terms.

L

Lanthanides – elements 57 (lanthanum) through 71 (lutetium); grouped because of their similar behavior in chemical reactions.

Lanthanide contraction – a decrease in the radii of the elements following the lanthanides compared to what would be expected if there were no *f*-transition metals.

Latent heat – the energy absorbed or released when a substance changes state without changing temperature.

Lattice – a unique arrangement of atoms or molecules in a crystalline liquid or solid.

Law of combining volumes (Gay-Lussac's Law) – at constant temperature and pressure, the volumes of reacting gases (and any gaseous products) can be expressed as ratios of small whole numbers.

Law of conservation of energy – energy cannot be created or destroyed; it can only change form.

Law of conservation of matter – there is no detectable change in the quantity of matter during an ordinary chemical reaction.

Law of conservation of matter and energy – the amount of matter and energy in the universe is fixed.

Law of definite proportions (law of constant composition) – different samples of a pure compound contain the same elements in the same proportions by mass.

Law of partial pressures (or *Dalton's Law*) – the pressure exerted by a mixture of gases is the sum of the partial pressures of the individual gases.

Laws of thermodynamics – physical laws which define quantities of thermodynamic systems describe how they behave and (by extension) set certain limitations, such as perpetual motion.

Lead storage battery – a secondary voltaic cell used in most automobiles.

Leclanche cell – a common type of *dry cell*.

Le Chatelier's principle – states that a system at equilibrium, or striving to attain equilibrium, responds in such a way as to counteract any stress placed upon it; if stress (change of conditions) is applied to a system at equilibrium, the system will shift in the direction that reduces stress.

Leveling effect – acids stronger than the acid characteristic of the solvent react with the solvent to produce that acid; a similar statement applies to bases. The strongest acid (base) that can exist in a given solvent is the acid (base) characteristic of the solvent.

Levorotatory (or *levo*) – an optically active substance rotates plane-polarized light counterclockwise.

Lewis acid – any species that can accept a share in an electron pair.

Lewis base – any species that can make available a pair of electrons.

Lewis dot formula (electron dot formula) – representation of a molecule, ion, or formula unit by showing atomic symbols and only outer shell electrons.

Ligand – a Lewis base in a coordination compound.

Light – that portion of the electromagnetic spectrum visible to the naked eye; known as "visible light."

Limiting reactant – a substance that stoichiometrically limits the number of products formed.

Linear accelerator – a device used for accelerating charged particles along a straight line path.

Line spectrum – an atomic emission or absorption spectrum.

Linkage isomers – a particular ligand bonds to a metal ion through different donor atoms.

Liquid – a state of matter which takes the shape of its container.

Liquid aerosol – colloidal suspension of a liquid in a gas.

London dispersion forces – very weak and very short-range attractive forces between short-lived temporary (induced) dipoles; known as "dispersion forces."

Lone pair – a pair of electrons residing on one atom and not shared by other atoms; an unshared pair.

Low spin complex – crystal field designation for an inner orbital complex; contains electrons paired t_{2g} orbitals before e_g orbitals are occupied in octahedral complexes.

Lubricant – a substance capable of reducing friction (i.e., a force that opposes the direction of motion).

M

Magnetic field – a space around a magnet where magnetism can be detected.

Magnetic quantum number – quantum mechanical solution to a wave equation designating the orbital within a given set (s, p, d, f) in which an electron resides.

Manometer – a two-armed barometer.

Mass (m) – a measure of the amount of matter in an object; mass is usually measured in grams or kilograms.

Mass action expression – for a reversible reaction, aA + bB cC + dD; the product of the concentrations of the products (species on the right), each raised to the power corresponding to its coefficient in the balanced chemical equation, divided by the product of the concentrations of reactants (species on the left), each raised to the power corresponding to its coefficient in the balanced equation; at equilibrium the mass action expression is equal to K.

Mass deficiency – the amount of matter converted into energy when an atom forms from constituent particles.

Mass number – the sum of the numbers of protons and neutrons in an atom; an integer.

Mass spectrometer – an instrument that measures the charge-to-mass ratio of charged particles.

Matter – anything that has mass and occupies space.

Mechanism – the sequence of steps by which reactants are converted into products.

Melting point – the temperature at which liquid and solid coexist in equilibrium.

Meniscus – the shape assumed by the surface of a liquid in a cylindrical container.

Melting – the phase change from a solid to a liquid.

Metal – a chemical element that is a good conductor of electricity and heat and forms cations and ionic bonds with nonmetals; elements below and to the left of the stepwise division (metalloids) in the upper right corner of the periodic table; about 80% of known elements are metals.

Metallic bonding – bonding within metals due to the electrical attraction of positively charged metal ions for mobile electrons that belong to the crystal.

Metallic conduction – conduction of electrical current through a metal or along a metallic surface.

Metalloid – a substance with the properties of metals and nonmetals (B, Al, Si, Ge, As, Sb, Te, Po and At).

Metathesis reactions – reactions in which two compounds react to form two new compounds, with no changes in oxidation number; reactions in which the ions of two compounds exchange partners.

Method of initial rates – method of determining the rate-law expression by carrying out a reaction with different initial concentrations and analyzing the resultant changes in initial rates.

Methylene blue – a heterocyclic aromatic chemical compound with the molecular formula $C_{16}H_{18}N_3SCl$.

Miscible liquids – mix to form a solution (e.g., alcohol and water).

Miscibility – the ability of one liquid to mix with (dissolve in) another liquid.

Mixture – two or more different substances mingled together but not chemically combined. A sample of matter composed of two or more substances, each of which retains its identity and properties.

Moderator – a substance (e.g., deuterium, oxygen, paraffin) that slows fast neutrons upon collision.

Molality – a concentration expressed as the number of moles of solute per kilogram of solvent.

Molarity – the number of moles of solute per liter of solution.

Molar solubility – the number of moles of a solute that dissolves to produce a liter of a saturated solution.

Mole – a measurement of an amount of substance; a single mole contains approximately 6.022×10^{23} units or entities; abbreviated mol.

Molecule – a chemically bonded number of electrically neutral atoms.

Molecular equation – a chemical reaction in which formulas are written as if substances existed as molecules; only complete formulas are used.

Molecular formula – indicates the number of atoms present in a molecule of a molecular substance.

Molecular geometry – the arrangement of atoms (not lone pairs of electrons) around a central atom of a molecule or polyatomic ion.

Molecular orbital (*MO*) – resulting from the overlap and mixing of atomic orbitals on different atoms (i.e., a region where an electron can be found in a molecule, as opposed to an atom); an MO belongs to the molecule.

Molecular orbital theory – a theory of chemical bonding based upon postulated molecular orbitals.

Molecular weight – the mass of one molecule of a nonionic substance in atomic mass units.

Molecule – the smallest particle of a compound capable of stable, independent existence.

Mole fraction – the number of moles of a component in a mixture divided by the number of moles in the mixture.

Monoprotic acid – can form only one hydronium ion per molecule; may be strong or weak.

Mother nuclide – nuclide that undergoes nuclear decay.

N

Native state – refers to the occurrence of an element in an uncombined or free state in nature.

Natural radioactivity – spontaneous decomposition of an atom.

Neat – conditions with a liquid reagent or gas performed with no added solvent or co-solvent.

Nernst equation – corrects standard electrode potentials for nonstandard conditions.

Net ionic equation – results from canceling spectator ions and eliminating brackets from a total ionic equation.

Neutralization – the reaction of an acid with a base to form a salt and water; usually, the reaction of hydrogen ions with hydroxide ions to form water molecules.

Neutrino – a particle that travels at speeds close to the speed of light; created due to radioactive decay.

Neutron – a neutral unit or subatomic particle with no net charge and a mass of 1.0087 amu.

Nickel-cadmium cell (NiCd battery) – a dry cell where the anode is Cd, the cathode is NiO_2, and the electrolyte is basic.

Nitrogen cycle – the complex series of reactions by which nitrogen is slowly but continually recycled in the atmosphere, lithosphere, and hydrosphere.

Noble gases – elements of the periodic Group 0; He, Ne, Ar, Kr, Xe, Rn; known as "rare gases;" formerly called "inert gases."

Nodal plane – a region in which the probability of finding an electron is zero.

Nonbonding orbital – a molecular orbital derived only from an atomic orbital of one atom; it lends neither stability nor instability to a molecule or ion when populated with electrons.

Nonelectrolyte – a substance whose aqueous solutions do not conduct electricity.

Nonmetal – an element that is not metallic.

Nonpolar bond – a covalent bond in which electron density is symmetrically distributed.

Nuclear – of or about the atomic nucleus.

Nuclear binding energy – the energy equivalent of the mass deficiency; the energy released in forming an atom from the subatomic particles.

Nuclear fission – when a heavy nucleus splits into nuclei of intermediate masses, and protons are emitted.

Nuclear magnetic resonance spectroscopy – a technique that exploits the magnetic properties of specific nuclei; helps identify unknown compounds.

Nuclear reaction – involves a change in the composition of a nucleus and can emit or absorb a tremendous amount of energy.

Nuclear reactor – a system in which controlled nuclear fission reactions generate heat energy on a large scale, subsequently converted into electrical energy.

Nucleons – particles comprising the nucleus: protons and neutrons.

Nucleus – the tiny and dense, positively charged center of an atom containing protons and neutrons, as well as other subatomic particles; the net charge is positive.

Nuclides – refers to different atomic forms of elements; in contrast to isotopes, which refer only to different atomic forms of a single element.

Nuclide symbol – designation for an atom A/Z E, in which E is the symbol of an element, Z is its atomic number, and A is its mass number.

Number density – a measure of the concentration of countable objects (e.g., atoms, molecules) in space; the number per volume.

O

Octahedral – molecules and polyatomic ions with one atom in the center and six atoms at the corners of an octahedron.

Octane number – a number that indicates how smoothly a gasoline burns.

Octet rule – during bonding, atoms tend to reach an electron arrangement with eight electrons in the outermost shell. Many representative elements attain at least a share of eight electrons in their valence shells when they form molecular or ionic compounds; there are some limitations.

Open sextet – species with only six electrons in the highest energy level of the central element (many Lewis acids).

Orbital – may refer to an atomic orbital or a molecular orbital.

Organic chemistry – the chemistry of substances that contain carbon-hydrogen bonds.

Organic compound – substances that contain carbon.

Osmosis – when solvent molecules pass through a semi-permeable membrane from a dilute solution into a more concentrated solution.

Osmotic pressure – the hydrostatic pressure produced on the surface of a semi-permeable membrane.

Outer orbital complex – valence bond designation for a complex in which the metal ion utilizes d orbitals in the outermost (occupied) shell in hybridization.

Overlap – the interaction of orbitals on different atoms in the same region of space.

Oxidation – the addition of oxygen or the loss of electrons. An algebraic increase in the oxidation number may correspond to a loss of electrons.

Oxidation numbers – quantitative values used as mechanical aids in writing formulas and balancing equations; for single-atom ions, they correspond to the charge on the ion; more electronegative atoms are assigned negative oxidation numbers, known as *oxidation states*.

Oxidation-reduction reactions – reactions in which oxidation and reduction occur; known as *redox reactions*.

Oxide – a binary compound of oxygen.

Oxidizing agent – the substance that oxidizes another substance and is reduced.

P

Pairing – a favorable interaction of two electrons with opposite m values in the same orbital.

Pairing energy – the energy required to pair two electrons in the same orbital.

Paramagnetism – attraction toward a magnetic field, stronger than diamagnetism but still weak compared to ferromagnetism.

Partial pressure – the force exerted by one gas in a mixture of gases.

Particulate matter – fine, divided solid particles suspended in polluted air.

Pauli exclusion principle – no electrons in the same atom may have identical sets of four quantum numbers.

Percentage ionization – the percentage of the weak electrolyte that will ionize in a solution of given concentration.

Percent by mass – 100% times the actual yield divided by the theoretical yield.

Percent composition – the mass percent of each element in a compound.

Percent purity – the percent of a specified compound or element in an impure sample.

Period – the elements in a horizontal row of the periodic table.

Periodicity – regular periodic variations of properties of elements with their atomic number (and position in the periodic table).

Periodic Law – the properties of the elements are periodic functions of their atomic numbers.

Periodic table – an arrangement of elements by increasing atomic numbers, emphasizing periodicity.

Peroxide – a compound with oxygen in –1 oxidation state; metal peroxides contain the peroxide ion, O_2^{2-}.

pH – the measure of acidity (or basicity) of a solution; negative logarithm of the concentration (mol/L) of the H_3O^+ [H^+] ion; scale is commonly used over a range 0 to 14.

Phase diagram – shows the equilibrium temperature-pressure relationships for different phases of a substance.

pH scale – a range from 0 to 14. If the pH of a solution is 7, it is neutral; if the pH of a solution is less than 7, it is acidic; if the pH of a solution is greater than 7, it is basic.

Permanent hardness – hardness (relative to lathering soap) in water that cannot be removed by boiling; caused by calcium sulfate.

Photoelectric effect – emission of an electron from the surface of a metal caused by impinging electromagnetic radiation of specific minimum energy; the current increases with increasing radiation intensity.

Photon – a carrier of electromagnetic radiation of all wavelengths, such as gamma rays and radio waves; known as a *quantum of light*.

Physical change – when a substance changes from one physical state to another, but no substances with different compositions are formed; physical change may involve a phase change (e.g., melting, freezing) or another physical change, such as crushing a crystal or separating one volume of liquid into different containers; it does not produce a new substance.

Plasma – a physical state of matter that exists at extremely high temperatures in which molecules are dissociated, and most atoms are ionized.

Polar bond – a covalent bond with an unsymmetrical distribution of electron density.

Polarimeter – a device used to measure optical activity.

Polarization – the buildup of a product of oxidation or reduction of an electrode, preventing further reaction.

Polydentate – refers to ligands with more than one donor atom.

Polyene – a compound that contains more than one double bond per molecule.

Polymerization – the combination of many small molecules to form large molecules.

Polymer – a large molecule consisting of chains or rings of linked monomer units, usually characterized by high melting and boiling points.

Polymorphous – refers to substances that can crystallize in more than one crystalline arrangement.

Polyprotic acid – forms two or more H_3O+ ions per molecule; often, at least one ionization step is weak.

Positron – a nuclear particle with the mass of an electron but opposite charge (positive).

Potential difference (or *voltage*) – the force that moves the electrons around circuit; unit is Volt (V).

Potential energy (*PE*) – energy stored in a body or a system due to its position in a force field or configuration.

Power – the rate at which energy is converted from one form to another; the unit is Watts (W). Power = voltage $\times$ current (P = VI).

Precipitate – an insoluble solid formed by mixing in solution the constituent ions of a slightly soluble solution.

Precision – how close the results of multiple experimental trials are; see *accuracy*.

Pressure – force per unit area; unit is Pascal (Pa).

Primary standard – a known high degree of purity substance that undergoes one invariable reaction with the other reactant of interest.

Primary voltaic cells – voltaic cells that cannot be recharged; no further chemical reaction is possible once the reactants are consumed.

Products – chemicals produced (from reactants) in a chemical reaction.

Proton – a subatomic particle having a mass of 1.0073 amu and a charge of $+1$, found atom's nucleus.

Protonation – the addition of a proton (H^+) to an atom, molecule, or ion.

Pseudobinaryionic compounds – contain more than two elements but are named like binary compounds.

Q

Quanta – the minimum amount of energy emitted by radiation.

Quantum mechanics – the study of how atoms, molecules, subatomic particles, etc., behave and are structured; a mathematical method of treating particles based on quantum theory, which assumes that energy (of small particles) is not infinitely divisible.

Quantum numbers – numbers that describe the energies of electrons in atoms; derived from quantum mechanical treatment.

Quarks – elementary particles and a fundamental constituent of matter, combining to form hadrons (i.e., protons and neutrons).

R

Radiation – 1) heat transfer through invisible rays, which travel outwards from the hot object without a medium. 2) high-energy particles or rays emitted during the nuclear decay processes.

Radical – an atom or group of atoms that contains one or more unpaired electrons; usually a very reactive species.

Radioactive dating – a method of dating ancient objects by determining the ratio of mother and daughter nuclides present in an object and relating the ratio to the object's age via half-life calculations.

Radioactive tracer – a small amount of radioisotope replacing a nonradioactive isotope of the element in a compound whose path (e.g., in the body) or whose decomposition products are monitored by detection of radioactivity; known as a "radioactive label."

Radioactivity – the spontaneous disintegration of atomic nuclei.

Raoult's Law – the vapor pressure of a solvent in an ideal solution decreases as its mole fraction decreases.

Rate-determining step – the slowest step in a mechanism; determines the overall reaction rate.

Rate-law expression – an equation relating the reaction rate to the concentrations of the reactants and the specific rate of the reaction.

Rate of reaction – the change in the concentration of a reactant or product per unit time.

Reactants – substances consumed in a chemical reaction; react together in a chemical reaction.

Reaction quotient – the mass action expression under any set of conditions (not necessarily equilibrium); its magnitude relative to K determines the direction in which the reaction must occur to establish equilibrium.

Reaction ratio – the relative amounts of reactants and products involved in a reaction; may be the ratio of moles, millimoles, or masses.

Reaction stoichiometry – describes the quantitative relationships among substances participating in chemical reactions.

Reactivity series (or activity series) – an empirical, calculated, and structurally analytical progression of a series of metals, arranged by "reactivity" from highest to lowest; used to summarize information about the reactions of metals with acids and water, double displacement reactions, and the extraction of metals from ores.

Reagent – a substance (or compound) added to a system to cause a chemical reaction or to visualize if a reaction occurs; the terms reactant and reagent are often used interchangeably; however, a *reactant* is more specifically a substance consumed during a chemical reaction.

Reducing agent – a substance that reduces another substance and is itself oxidized.

Reduction – the removal of oxygen or the gaining of electrons.

Resonance – the concept in which two or more equivalent dot formulas for the same arrangement of atoms (resonance structures) are necessary to describe the bonding in a molecule or ion.

Reverse osmosis – forcing solvent molecules to flow through a semi-permeable membrane from a concentrated solution into a dilute solution by applying greater hydrostatic pressure on the concentrated side than the osmotic pressure opposing it.

Reversible reaction – processes that do not go to completion and occur in the forward and reverse direction.

S

Saline solution – a general term for NaCl (i.e., sodium chloride) in water.

Salt – when a metal replaces the hydrogen of an acid.

Salts – ionic compounds composed of anions and cations.

Salt bridge – a U-shaped tube containing an electrolyte, connects the two half-cells of a voltaic cell.

Saturated solution –no more solute will dissolve at that temperature.

s-**block elements** – group 1 and 2 elements (alkali and alkaline metals), including hydrogen and helium.

Schrödinger equation – quantum state equation representing the behavior of an electron around an atom; describes the wave function of a physical system evolving.

Second Law of Thermodynamics – the universe tends toward a state of greater disorder in spontaneous processes.

Secondary standard – a solution that has been titrated against a primary standard; a standard solution.

Secondary voltaic cells – voltaic cells that can be recharged; original reactants can be regenerated by reversing the direction of the current flow.

Semiconductor – a substance that does not conduct electricity at low temperatures but will do so at higher temperatures.

Semi-permeable membrane – a thin partition between two solutions through which specific molecules can pass but others cannot.

Shielding effect – electrons in filled sets of *s, p* orbitals between the nucleus and outer shell electrons shield the outer shell electrons somewhat from the effect of protons in the nucleus; known as the "screening effect."

Sigma (σ) bonds – bonds resulting from the head-on overlap of atomic orbitals. The region of electron sharing is along and (cylindrically) symmetrical to the imaginary line connecting the bonded atoms.

Sigma orbital – molecular orbital resulting from the head-on overlap of two atomic orbitals.

Single bond – covalent bond resulting from the sharing of two electrons (one pair) between two atoms.

Sol – a suspension of solid particles in a liquid; artificial examples include sol-gels.

Solid – one of the states of matter, where the molecules are packed closely, resistance to movement/deformation, and volume change.

Solubility product constant – equilibrium constant for the dissolution of a slightly soluble compound.

Solubility product principle – the solubility product constant expression for a slightly soluble compound is the product of the concentrations of the constituent ions; each raised to the power that corresponds to the number of ions in one formula unit.

Solute – the dispersed (i.e., dissolved) phase of a solution; the solution is mixed into the solvent (e.g., NaCl in saline water).

Solution – a homogeneous mixture of multiple substances; comprised of solutes and solvents; a mixture of a solute (usually a solid) and a solvent (usually a liquid).

Solvation – the process by which solvent molecules surround and interact with solute ions or molecules.

Solvent – the dispersing medium of a solution (e.g., H_2O in saline water).

Solvolysis – the reaction of a substance with the solvent in which it is dissolved.

***s*-orbital** – a spherically symmetrical atomic orbital; one per energy level.

Specific gravity – the ratio of the density of a substance to the density of water.

Specific heat – the amount of heat required to raise the temperature of one gram of substance 1 °C.

Specific rate constant – an experimentally determined (proportionality) constant; different for different reactions, and which changes only with temperature; k in the rate-law expression: Rate = k [A] × [B].

Spectator ions – ions in a solution that do not participate in a chemical reaction.

Spectral line – any of several lines corresponding to definite wavelengths of an atomic emission or absorption spectrum; marks the energy difference between two energy levels.

Spectrochemical series – arrangement of ligands in order of increasing ligand field strength.

Spectroscopy – the study of radiation and matter, such as X-ray absorption and emission spectroscopy.

Spectrum – display of component wavelengths (colors) of electromagnetic radiation.

Speed of light – the speed at which radiation travels through a vacuum (299,792,458 m/sec).

Square planar – describes molecules and polyatomic ions with one atom in the center and four atoms at the corners of a square.

Square planar complex – relationship with metal in the center of a square plane, with ligand donor atoms at each of the four corners.

Standard conditions for temperature and pressure (STP) – used to compare experimental results (25 °C and 100.000 kPa).

Standard electrodes – half-cells in which the oxidized and reduced forms of a species are present at the unit activity (1.0 M solutions of dissolved ions, 1.0 atm partial pressure of gases, pure solids, and liquids).

Standard electrode potential – by convention, the potential (Eo) of a half-reaction as a reduction relative to the standard hydrogen electrode when species are present at unit activity.

Standard entropy – the absolute entropy of a substance in its standard state at 298 K.

Standard molar enthalpy of formation – the amount of heat absorbed in forming one mole of a substance in a specified state from its elements in their standard states.

Standard molar volume – space occupied by 1 mole of ideal gas under standard conditions; 22.4 liters.

Standard reaction – a process where the numbers of moles of reactants in the balanced equation, in their standard states, are entirely converted to the numbers of moles of products in the balanced equation, at their standard state.

State of matter – a homogeneous, macroscopic phase (e.g., gas, plasma, liquid, solid) in increasing concentration.

Stoichiometry – quantitative relationships of elements and compounds undergoing chemical changes.

Strong electrolyte – a substance that conducts electricity well in a dilute aqueous solution.

Strong field ligand – a ligand that exerts a strong crystal or ligand electrical field and generally forms low-spin complexes with metal ions when possible.

Structural isomers – compounds that contain the same number and kinds of atoms, but with different geometries.

Subatomic particles – comprise an atom (e.g., protons, neutrons, electrons).

Sublimation – the direct vaporization of a solid by heating without passing through the liquid state; a phase transition from solid to gas.

Substance – any matter, specimens with the same chemical composition and physical properties.

Substitution reaction – a reaction in which another atom or group of atoms replaces an atom or a group of atoms.

Supercooled liquids – liquids that, when cooled, apparently solidify but continue to flow very slowly under the influence of gravity.

Supercritical fluid – a substance at a temperature above its critical temperature.

Supersaturated solution – contains a higher than saturation concentration of solute; slight disturbance or seeding causes crystallization of excess solute.

Suspension – a heterogeneous mixture in which solute-like particles settle out of the solvent-like phase sometime after their introduction. A mixture of a liquid and a finely divided insoluble solid.

T

Talc – a mineral representing the Mohs Scale, composed of hydrated magnesium silicate with the chemical formula $H_2Mg_3(SiO_3)_4$ or $Mg_3Si_4O_{10}(OH)_2$.

Temperature – a measure of heat intensity (i.e., hotness or coldness of a sample); a measure of kinetic energy of an object. Units are measured in degrees, and scales include Celsius, Fahrenheit, and Kelvin.

Temporary hardness – hardness in water, removed by boiling; caused by calcium hydrogen carbonate.

Ternary acid – a ternary compound containing H, O and another element, often a nonmetal.

Ternary compound – a compound consisting of three elements; may be ionic or covalent.

Tetrahedral – a term used to describe molecules and polyatomic ions with one atom in the center and four atoms at the corners of a tetrahedron.

Theoretical yield – the maximum amount of a specified product that could be obtained from specified amounts of reactants, assuming complete consumption of the limiting reactant according to only one reaction and complete recovery of the product; see *actual yield*.

Theory – a model describing the nature of a phenomenon.

Thermal conductivity – a property of a material to conduct heat (often noted as k).

Thermal cracking – decomposition by heating a substance in the presence of a catalyst and without air.

Thermochemistry – the study of absorption/release of heat within a chemical reaction; studies heat energy associated with chemical reactions and physical transformations.

Thermodynamics – studying the effects of changing temperature, volume, or pressure (or work, heat, and energy) on a macroscopic scale.

Thermodynamic stability – when a system is in its lowest energy state with its environment (equilibrium).

Thermometer – a device that measures the average energy of a system.

Thermonuclear energy – energy from nuclear fusion reactions.

Third Law of Thermodynamics – entropy of a pure crystalline substance at absolute zero equals zero.

Titration – a procedure in which one solution is added to another solution until the chemical reaction between the two solutions is complete; the concentration of one solution is known, and that of the other is unknown. The process of adding one solution to a measured amount of another to find out exactly how much of each is required to react.

Torr – a unit to measure pressure; 1 Torr is equivalent to 133.322 Pa or 1.3158×10^{-3} atm.

Total ionic equation – the expression for a chemical reaction written to show the predominant form of species in aqueous solution or contact with water.

Transition elements (metals) – B Group elements except IIB in the periodic table; sometimes called transition elements, elements with incomplete *d* sub-shells; the *d*-block elements.

Transition state theory –reactants pass through high-energy transition states before forming products.

Transuranic element – an atomic number greater than 92; none of the transuranic elements are stable.

Triple bond – the sharing of three pairs of electrons within a covalent bond (e.g., N_2).

Triple point – where the temperature and pressure of three phases are the same; water has a unique phase diagram.

Tyndall effect – results from light scattering by colloidal particles (a mixture where one substance is dispersed evenly throughout another) or by suspended particles.

U

Uncertainty –any measurement that involves estimating any amount that cannot be precisely reproduced.

Uncertainty principle – knowing the location of a particle makes the momentum uncertain, while knowing the momentum of a particle makes the location uncertain.

Unit cell – the smallest repeating unit of a lattice.

Unit factor – statements used in converting between units.

Universal (or ideal) gas constant – proportionality constant in the ideal gas law ($0.08206 \ \text{L} \cdot \text{atm}/(\text{K} \cdot \text{mol})$).

UN number – a four-digit code used to note hazardous and flammable substances.

Unsaturated hydrocarbons – hydrocarbons that contain double or triple carbon-carbon bonds.

V

Valence bond theory – proposes that covalent bonds are formed when atomic orbitals on different atoms overlap and the electrons are shared.

Valence electrons – outermost electrons of atoms; usually those involved in bonding.

Valence shell electron pair repulsion theory (VSEPR) – assumes electron pairs are arranged around the central element of a molecule or polyatomic ion with maximum separation (and minimum repulsion) among regions of high electron density.

Valency – the number of electrons an atom wants to gain, lose, or share to have a full outer shell.

Van der Waals' equation – a quantitative relationship of a state extending the ideal gas law to real gases by including two empirically determined parameters, which are specific for different gases.

Van der Waals force – one of the forces (attraction/repulsion) between molecules.

Van't Hoff factor – the ratio of moles of particles in solution to moles of solute dissolved.

Vapor – when a substance is below the critical temperature in the gas phase.

Vaporization – the phase change from liquid to gas.

Vapor pressure – the particle pressure of vapor at the surface of its parent liquid.

Viscosity – the resistance of a liquid to flow (e.g., oil has a higher viscosity than water).

Volt – one joule of work per coulomb; the unit of electrical potential transferred.

Voltage – the potential difference between two electrodes; a measure of the chemical potential for a redox reaction.

Voltaic cells – electrochemical cells in which spontaneous chemical reactions produce electricity; known as *galvanic cells*.

Voltmeter – an instrument that measures the cell potential.

Volumetric analysis – measuring the volume of a solution (of known concentration) to determine the substance's concentration within the solution; see *titration*.

W

Water equivalent – the amount of water absorbing the same heat as the calorimeter per degree of temperature increase.

Weak electrolyte – a substance that conducts electricity poorly in a dilute aqueous solution.

Weak field ligand – a ligand that exerts a weak crystal or ligand field and generally forms high-spin complexes with metals.

X

X-ray – electromagnetic radiation between gamma and UV rays.

X-ray diffraction – a method for establishing structures of crystalline solids using single-wavelength X-rays and studying the diffraction pattern.

X-ray photoelectron spectroscopy – a spectroscopic technique used to measure the composition of a material.

Y

Yield – the amount of product produced during a chemical reaction.

Z

Zone melting – remove impurities from an element by melting and slowly traveling it down an ingot (cast).

Zone refining – a method of purifying a metal bar by passing it through an induction heater; this causes impurities to move along a melted portion.

Zwitterion (formerly called a dipolar ion) – a neutral molecule with a positive and negative electrical charge; multiple positive and negative charges can be present, distinct from dipoles at different locations within that molecule; known as *inner salts*.

Frank J. Addivinola, Ph.D.

This study guide's lead author and chief editor is Dr. Frank Addivinola. With his outstanding education, laboratory research, and decades of university science teaching, Dr. Addivinola lent his expertise to oversee the development of this series.

During his extensive career, Dr. Addivinola held faculty positions at colleges and universities, including Harvard University, Johns Hopkins University, University of Maryland, and Northeastern University, and taught undergraduate and graduate-level courses in biology, biochemistry, organic chemistry, inorganic chemistry, physics, anatomy and physiology, medical terminology, nutrition, and medical ethics. He received several awards for his research and presentations.

Dr. Frank Addivinola conducted original research in developmental biology as a doctoral candidate and pre-IRTA fellow in Molecular and Cell Biology at the National Institutes of Health (NIH). His dissertation advisor was Nobel laureate Marshall W. Nirenberg, Chief of the Biochemical Genetics Laboratory at the National Heart, Lung, and Blood Institute (NHLBI). Before NIH, Dr. Addivinola researched prostate cancer in the Cell Growth and Regulation Laboratory of Dr. Arthur Pardee at the Dana Farber Cancer Institute of Harvard Medical School.

Dr. Addivinola holds an undergraduate degree in biology from Williams College. He completed his Masters at Harvard University, Masters in Biotechnology at Johns Hopkins University, and five other graduate degrees at the University of Maryland University College, Suffolk University, and Northeastern University.

Essential Chemistry Self-Teaching Guides

Electronic Structure & Periodic Table

Chemical Bonding

States of Matter & Phase Equilibria

Stoichiometry

Solution Chemistry

Chemical Kinetics & Equilibrium

Acids & Bases

Chemical Thermodynamics

Electrochemistry

Visit our Amazon store

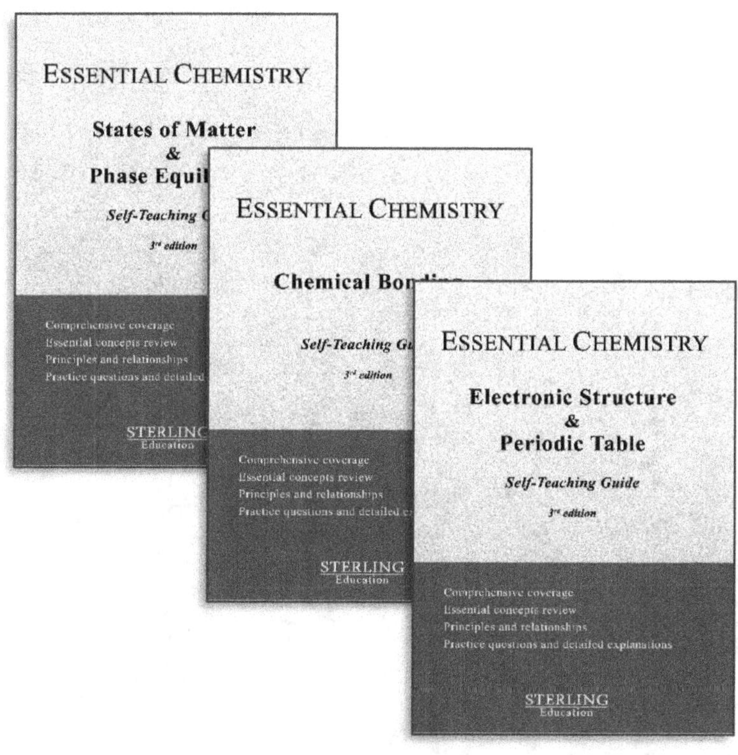

Essential Biology Self-Teaching Guides

Eukaryotic Cell & Cellular Metabolism

Molecular Biology & Genetics

Nervous & Endocrine Systems

Circulatory, Respiratory & Immune Systems

Digestive & Excretory Systems

Muscle, Skeletal & Integumentary Systems

Reproduction & Development

Microbiology

Plants & Photosynthesis

Evolution, Classification & Diversity

Ecology & Population Biology

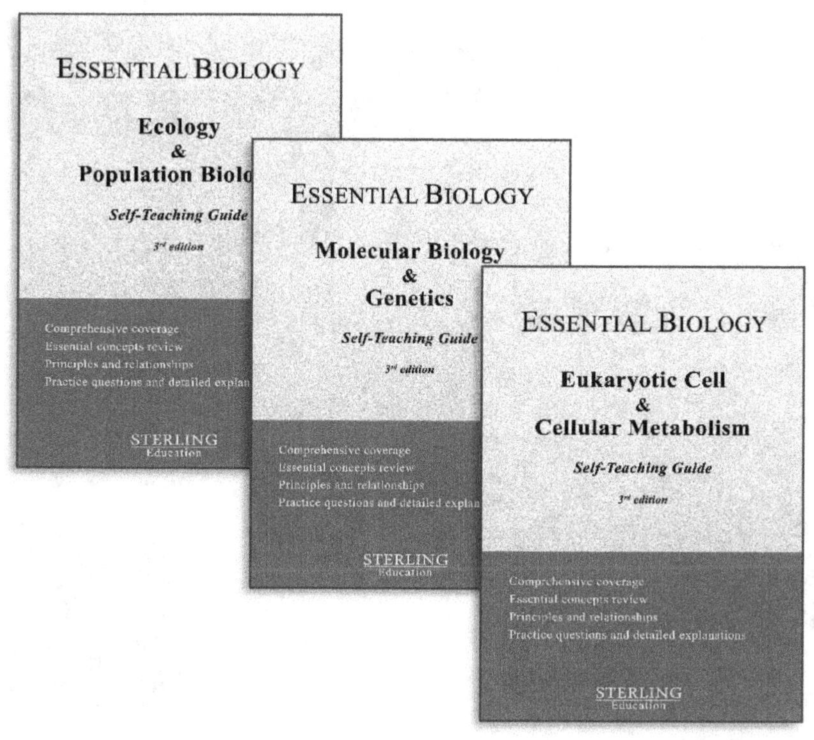

For online practice resources visit

https://www.sterling-prep.com